Jack Fieldhouse
Artist and Beekeeper

A visual journey

Artist and Beekeeper *A visual journey*
© Jack Fieldhouse

ISBN 978-1-908904-11-9

Published by Northern Bee Books, 2012
Scout Bottom Farm
Mytholmroyd
Hebden Bridge
HX7 5JS (UK)

Design and artwork
D&P Design and Print
Worcestershire

Printed by Lightning Source, UK

Jack Fieldhouse
Artist and Beekeeper
A *visual journey*

This book is dedicated to my friend
David Charles

Northern Bee Books

Preface

For almost all of our history since we emerged as a distinct species som 250,000 years ago, *Homo sapiens*, has made a living as a hunter-gathere in small bands. That all changed with the dawn of agriculture abou 10,000 years ago.

Agriculture had profound consequences for our subsequent histor With more food to go round, the human population increase inexorably from the estimated 1 million hunter-gatherers to 1 billion the beginning of the 19th century, nearly 7 billion today and a projecte 9 billion by 2050. Agriculture also resulted in a increase in wealth c some individuals and in the emergence of different roles in society farmer, soldier, merchant and so on.

Today, in rich countries such as the UK, we have unprecedente wealth and we consume more 'stuff than ever before. This combinatio growing population and increased consumption is putting intolerabl pressure on our environment. It has been estimated that everyone i the world consumed at the level of the USA, this would be equivalei to a global population of over 70 billion people. It would be an optimi indeed who would claim that the planet could support this number.

There are many indicators of our impact on the environmen induding loss of biodiversity and habitats, pollution, and degradatio of soils. But overarching all of these is climate change. Scientists agrt that by burning fossil fuels we are changing the composition of tt earth's atmosphere and this is causing climate change. If nothing is dor to halt this trend, many parts of the planet could become uninhabitab during the next few generations.

There are two bits of good news: first that environmental issue especially global warming, are gaining traction amongst politicians an decision makers, and second, that there are technological solution such as solar and wind energy, that will enable us to tackle the probler But it will also require political will and the support of all of us.

At the same time, we should all, as individuals, be aiming to reduce our environmental footprint and move to a more sustainable lifestyle. Bee-keeping is an ancient tradition that is well documented from Egypt and Middle East 4,000 years ago. And before this, hunter-gatherers collected honey from wild bees' nests. Even today, the Boran people of East Africa use a remarkable method to locate wild nests. A bird called the Greater Honey Guide feeds on the wax and larvae in a bees' nest, and it uses special calls to attract humans (and Honey Badgers) to a nest. Once the nest has been opened the birds feed on the discarded wax whilst the Boran people save the honey. Boran deploy a special loud whistle to call Honey Guides at the start of a nest-hunting trip. Today, beekeeping is widespread, with many millions of hives in Africa and Asia. It is an excellent example of a sustainable way of relating to the environment and providing food.

Lord Krebs, Principal, Jesus College Oxford.

Addendum

As we enter the second decade of the 21St century, there are encouraging signs that many people are interested in the countryside and all that it has to offer. The Royal Society for Protection of Birds has over 1 million members, the National Trust more than 2.5 million, and many smaller, local organisations engage tens of thousands of people with their interest in the natural environment and locally grown food. Bee-keeping is no exception. With the support of so many citizens, let us hope that the Government takes the environment seriously when formulating policy.

Bibliography
The Life of Jack Fieldhouse - Aged 92

Born in Sheffield (Attercliffe) in a house/one of a row of soot-covered dwellings in Clifford Street. Demolished in the 40s.

Father was chief Metallurgical Chemist at English Steels: Mother an artist.

Moved to Rotherham where we lived for a few years at Sherwood Crescent off Wellgate.

Attended school (St. Bedes Elementary) where I was taught by Miss Greaves who had also taught my father.

We were taught the three R's in a large schoolroom with a stove on which sat a large black kettle and this seemed to be permanently on the boil to provide Miss with her tea. It was surrounded by an iron fender to prevent accidents.

The furniture was of iron and wood and there was absolutely no noise.

High windows - no distraction. At eleven I went with my brother Bob to De La College at Pitsmoor in Sheffield.

In the winter it was a train journey, in the Summer I used my bicycle and a great joy it was to go at great speed down those Sheffield hills.

Art, English and Physics I enjoyed especially the Art but developed a loathing for Maths and to some extent History.

Left school at fourteen and decided to be a decorator, joining the firm of Banner-Cox in Percy Street, Rotherham.

As an apprentice I pushed a handcart in Rotherham or outlying village where I cleaned and painted gutters. Became adept at using ladders and one of my specialities was painting outside 'lav' doors. Plenty of in-door work too and I was paid eight shillings and sixpence a week. Overtime an extra sixpence an hour.

With war coming I convinced some friends and my brother that if we wanted to stay together we should join the local Territorial Unit at Wentworth Woodhouse – an Ack. Ack. Battery in the making.

Wentworth I might add was my favourite stamping ground in my teens - camping and bird watching and bathing in the lakes.

Also I had worked in the house and on one occasion her Ladyship took me into a large gallery to tell me about the portraits and artists.

In the autumn of 38 we were called up and sent to Hedon Airfield near Hull: It was one of those deep snow winters, unforgettably cold and unsuitable rations – I remember the large tins of pilchards which my mining colleague efused to eat and they built them into one gun emplacements to show their isapproval.

Local farms complained about missing chickens!

The following Spring we were moved to Godhill on the side opposite Hull nd manned heavy 4.5 Anti-Aircraft guns. 1940 saw a lot of bombing of Hull nd Immingham and we were constantly in action.

The following Winter was awful – bitterly cold and we were at stand-to om ten o'clock till four in the morning. Gerry was obviously annoyed by our ck. Ack. guns and tried to eliminate us by dropping a landmine by parachute hich fell into a ditch about a hundred yards away.

The H.Q. was in a farmhouse on the opposite side of the road – it was sliced a two and there were many casualties.

One gun crew on the Hull side took a direct hit and all were killed. We were 1oved about the country and had our 3.7 guns.

A fortnight's relief came when we were all moved to Burrow Head in Vigtownshire for 'Practise' where a plane towing a target appeared and we had o prove we could quickly react and destroy it. I didn't envy the pilot!

Wonderful place for rabbits and hare and as I carried a shotgun I provided change of diet for my comrades.

In 42 we were shipped to N. Africa, I took with me a pair of dark glasses, alcum powder, French and German text books, fishing line and several hooks or catching trout and all were used, especially the dark glasses and talcum owder in Africa.

I had some 'school' French and improved my knowledge of the language as 1ost of the Algerians I met spoke French.

Stationed at Sousse, South of Tunis, the C.O. decided to send me on a 0urse mostly to do with aircraft reorganisation (I was Lance Bombardier I/C f lecturing on aircraft recognition - unpaid). I had to go several hundreds of uiles back to Algeria for a week's cramming. I travelled in a cattle wagon with 1me American 'buddies' and on the way got a nasty dose of Delhi Belly or quits (such a descriptive word).

We trundled along and were pushed into a sidings at night.

My condition was perilous and there were no 'facilities' and was so weak 1d not wanting to embarrass anyone I stuck my backside out as far as I could, holding on to the ironwork with what strength I could muster.

As the squitty phase was over I lay on the wooden floor to recuperate unt the next 'bout.' I think that within a day I was back to normal, nasty though!

Saw the whole of Italy from South to North. Italy was beautiful, but ravage and as the Luftwaffe reduced we were transferred to Infantry (The Yorkshir Light Infantry).

Just before that I had a vivid memory of Cassino where we had a brie deployment and had to move out.

The noise was appalling with bombardment on the lower slopes an phalanges of American bombers reducing the Monastery to rubble.

The Poles and New Zealanders suffered heavy casualties and I remembe spending the day loading up into lorry, the 'regimental' ammunition. My frien Bombardier Kendrick and I were chosen for that job because we were 'stron lads'!

Later in training in the Apennines I was drafted in a Mule Corps and wit five comrades spent the winter and spring taking ammunition to the troops i the mountains.

The inhabitants of the village in which we were billeted were extreme poor. I was invited into a Dara where they lived. It had an earth floor, a stov for cooking, fuelled with wood and a huge bed in which the whole family slep end to end. One wall was taken up with wood collected to fuel the fire.

A bonus for them was the manure for their gardens and I remember ho four of us lifted a ground sheet containing at least a cwt of manure (and we tied) onto the back of a local woman.

How far away was the garden I knew not, but she was a one tough female

At the end of the war we moved into Unter Steimark next to the Yugosl border where the people there had been under Russian occupation.

It was the time when a very wrong Political Agreement was made to han back to the Russians, captured White Russians who had been conscripted in the German Army. There were many suicides

Onto Graz and Murguschlag where from we moved to Mariazell in th mountains. From Murguschlag (the Pit boys quickly renamed it the 'Sl Heap'). I was granted a three day pass to Vienna where I found people on th verge of starvation and our billet was an old hospital where the Russians ha moved out.

Their pallicses were stacked in corridors and our iron beds in a huge ward were doused in D.D.T.

In spite of this anti-bug precaution I was (as were we all) subjected to the tormenting bites of bed bugs.

People sat in cafes with nothing to drink. I think they were trying to keep up the spirit of Vienna.

Long queues for a meagre ration of milk and bread.

It took three days of careful scrutiny to rid my blanket of these blood suckers.

The condition of the average Russian soldier? I can only guess.

I had a stay at Gollrad one side of a mountain pass and I did the cooking on huge wood fuelled stove.

My 'forte' in the kitchen was dumplings boiled in socks! Very expansive!

Demob – over the mountain pass on a 'track' vehicle and eventually on to a most uncomfortable French train to one of the ports.

Thought twice about going back to Rotherham and I made up my mind that I would return when there were salmon in the Don.

I keep returning to Rotherham when staying with my architect brother Bob who lived at Tickhill near Doncaster and looking over the Chantry Bridge in Rotherham I saw a man fly fishing.

I settled in Woking with an aunt and managed to convince the Head of the Art School in Guildford that I had qualifications which could be worked on and he accepted.

Three years of training to obtain a degree in Fine Art and whilst there, volunteered to pick apples in Cambridge. My wife to be Joan had decided on the same course and we quickly became 'romantically involved'. We were married at Christmas: My aunt owned a property in Woking and we moved in.

It was a time when we were still on an egg a week ration: I bought three pullets and quite soon we were enjoying three eggs a day. I went out shooting rabbits and pigeons and we lived quite well.

After a year at London University I took a post of Art Master at a large secondary Modern. It was a poor school. The crème de la crème of the girls worked in Woolworth's, the 'A' stream were ok – they actually listened!

After five years we moved to Bridgwater and into the old house where w
still live. The school was boys only and on reflection slightly better. We ha
four children by then and another three arrived.

Five years later I moved to Priory Boys School in Taunton, again a Secondar
Modern but far superior to the ones in which I had spent the previous te
years. Educating six of my children privately drained us and holidays happene
by exchanging houses with friends who were having to cope with the sam
demands.

In the summer holidays with other members of staff we took boys t
Colonsay or Visr – memorable experiences for all and the family came too.

Lord and Lady Strathcona forbade camping but as I knew someone wh
knew them we got permission to camp on the lawn at Colonsay House and th
kitchen window was left permanently open for hot water!

It was here that I heard a sound that I hadn't heard since 1939 on th
Humber – a corncrake. A brilliant re-connection and it happened where th
mile wide strand is between Colonsey and Oronsay.

Teaching in 'the good old days' allowed me to slipper boys who stepped ove
the line but I preferred a flat stick and would tell the offensive one to come t
the front of the class and bend down and face Mecca (where's Mecca Zur.
And having given the one a map reference I would whack him on the backsic
and sat him down to put a notch in the side of the stick with my knife and te
them how the old gun slingers of the west put a notch on the handle of the
guns after having shot dead an opponent.

At 60 I decided I had had enough and asked the staff to give me as a leavir
present, a sledge hammer and a wedge which was particularly useful during tl
Dutch Elm disease phase as Elms are notoriously knotty and difficult to spil

31 years of retirement full of beekeeping problems, chores and ART.

Can recommend a few hundred stings to keep the rheums at bay!

And I read of Russian beekeepers who lived to well over a hundred becau:
they sold their refined honey but lived on the 'rough stuff' it being honey, wa
pollen and propolis!

Rough shooting with my dear friend Frank Morris I/C accounts at Coun
Hall.

We had brilliant pigeon shoots with an almost out-of-control spani orrowed from my farmer friend Ken Webber.

The 'children' were a joy and six of them professionals.

Have turned out a Nun now doing social work, a Vet married to a Vet, a acher of Autistics – I call her 'de fortibus' (strong minded) two Artists and Mechanic who deals with electronics and I ask myself 'where did that come om?'

My damaged boy Duncan can communicate a little, he is very strong and fit nd does jobs around the house mostly the heavy work.

Goes for long walks and brings back logs or stumps of trees. Quite recently ne evening it was almost dark and he had been out over six hours and so I ecided to take the van and search. I found him coming into the lane with a uge log of wood and another one which he couldn't carry being pushed along y foot.

It would have taken him another half an hour to roll it home, but it does low how determined he is.

Grandchildren of course always welcome if they have good appetites. Ianners we take for granted.

Conclude this story in Mid Wales in am old Elizabethan house with stone oors and dark oak carvings and later in the day to an Art Exhibition at the ishops Palace, Hereford where my daughter Julie is exhibiting.

Another delightful moment!

Jack Fieldhouse
For Hall in Powys.

1938

During the Summer of 1938, my two friends Geof Lowe and Philip Smith bought an old lifeboat. It had been badly repaired and was towed along the canal from the Humber to Rotterham where it was beached.

At that time I was working for a decorator, a certain Banner Cox and he allowed me to have an amount of lead and zinc paint with which we gave the boat two coats.

This after repairs of course and the whole Summer seemed to be taken up preparing the lifeboat for a new phase of existence.

In 1939 with the war approaching I was called in the colours and departed with my Anti Aircraft Battery to the Humber where we saw action (and suffered too many sleepless nights). Action was a relief, it was the long "standing to" for months and usually up at four in the morning that was the real trial. Never have I been so cold.

Bed for three hours and then up to do a route march!
Times were tough!

I met Geof Lowe on leave and he told me that the lifeboat had been birthed near Hull and when our gallant army had to retreat through Dunkirk, the lifeboat was commandeered and brought soldiers from the beaches. The summer of 1938 had not been a waste of time!

Jack Fieldhou

(Ex Artillery and Infantry)

1944

At the end of the war in Italy my unit of the Kings Own Yorkshire Light Infantry moved into Unter Steimark a part of Austria next to the Balkans.

The Russians moved out, they had been occupying the land during the war. It was beautiful country and those who were left were impoverished. After a couple of days and after settling in, I had time to explore and with a stick set off to look at a local village.

On entering the wide street there was no-one in sight and it seemed completely deserted. Then I saw a face and then another and then two children who had been watching me came out, cautiously at first, then another and suddenly more and quite quickly I was surrounded by women and children.

No men, they had all been taken away to work in Russia. A silence descended and then a girl asked in English "Where are your guns?" Feeling rather perplexed I said that I had no guns, only a stick. This was the first time they had seen an Englishman! A contrast indeed to the heavily armed Russians.

They were intensely relieved and completely encircled me to ask questions. They had nothing except their bedraggled clothes, their simple houses and gardens.

They were self-sufficient and when at a later date I cycled to them from Klagenfurt to see them, they fed me on a huge slice of blackbread literally covered with lard.

The moment of meeting is always with me with a sense of thankfulness that I was privileged to meet such stoic people and to give them a slight ray of hope for the future.

I'm sure Herr Jack has been talked about for at least two generations! The question is still with me – "when if ever would they see their menfolk?"

Jack Fieldhouse

Aged 87 and written at Broomfield Hall, Enmore, Bridgewater

May 2007

Tanging the Bees

June, and a swarm has issued. Mother holds a pan and beats it with a ladle to make the bees cluster and to attract the attention of father in the hayfield beyond.

The swarm has settled on the pig hook.

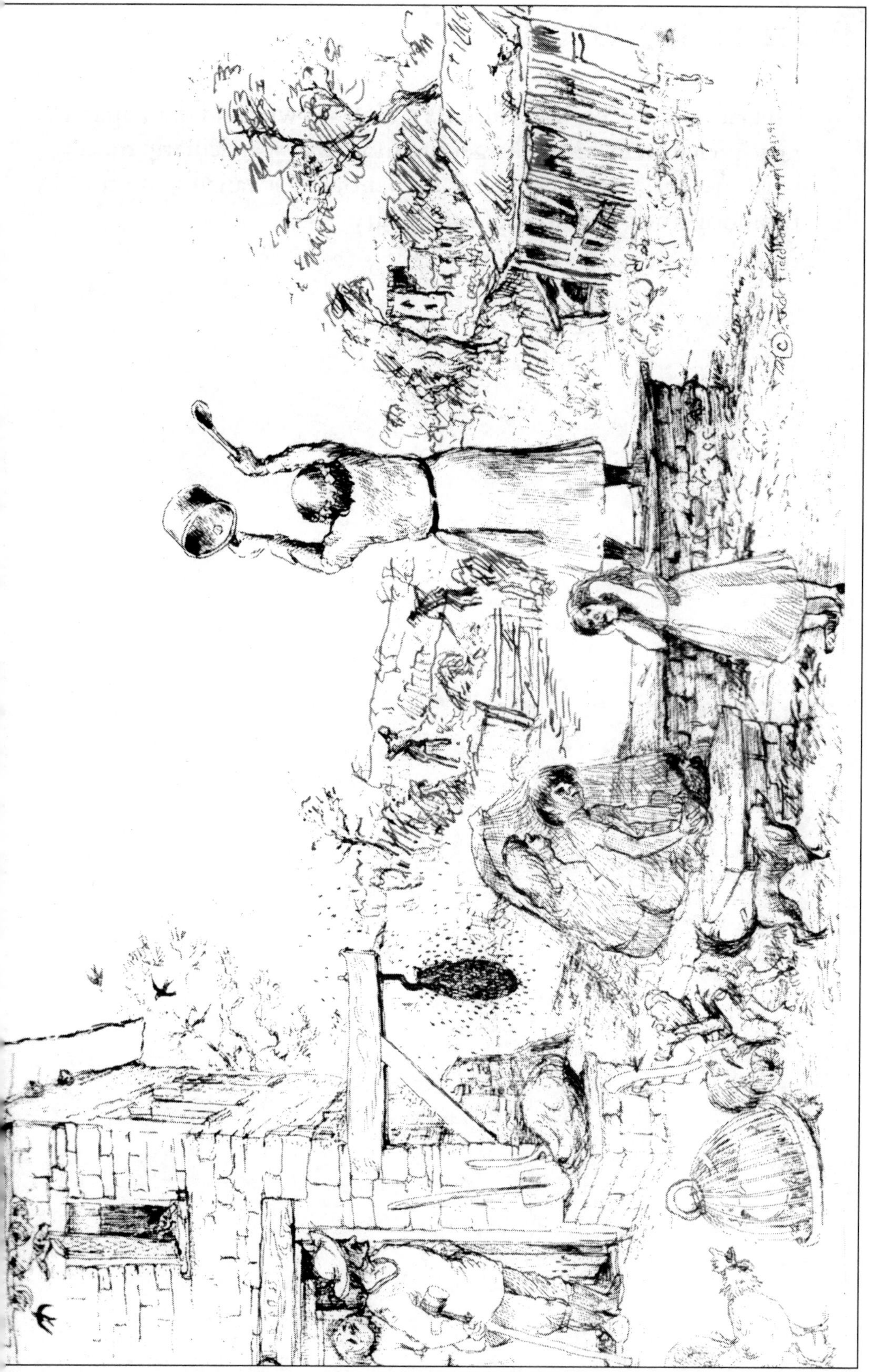

'Admiring the Bees'

The Reverend Charles Philpotts shows his wife and daughters his new 'Neighbour' hive. The gentry and clergy, particularly those with a 'scientific aptitude', were influential in advancing the craft by promoting new methods of husbandry.

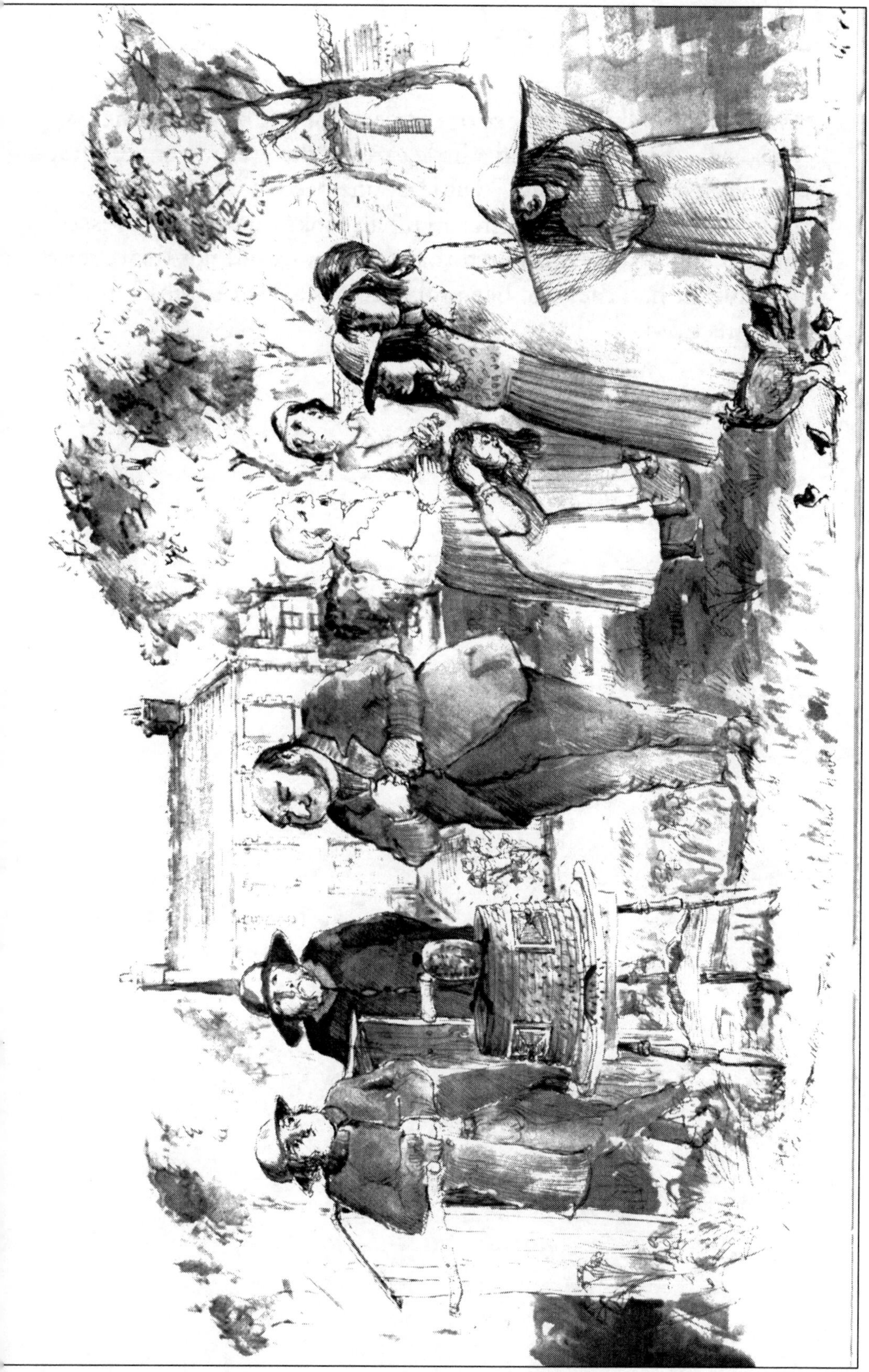

'Telling the Bees'

Grandma is telling the bees of the death of her husband, the beekeeper. She has arrived a little early in the evening and in draping the black crepe has annoyed at least one bee which flew out and lodged in Maude's hair. Sister Hannah looks on with suppressed merriment. It was considered that the bees would not thrive for their new owner had they not been told. This is a custom which is still practiced.

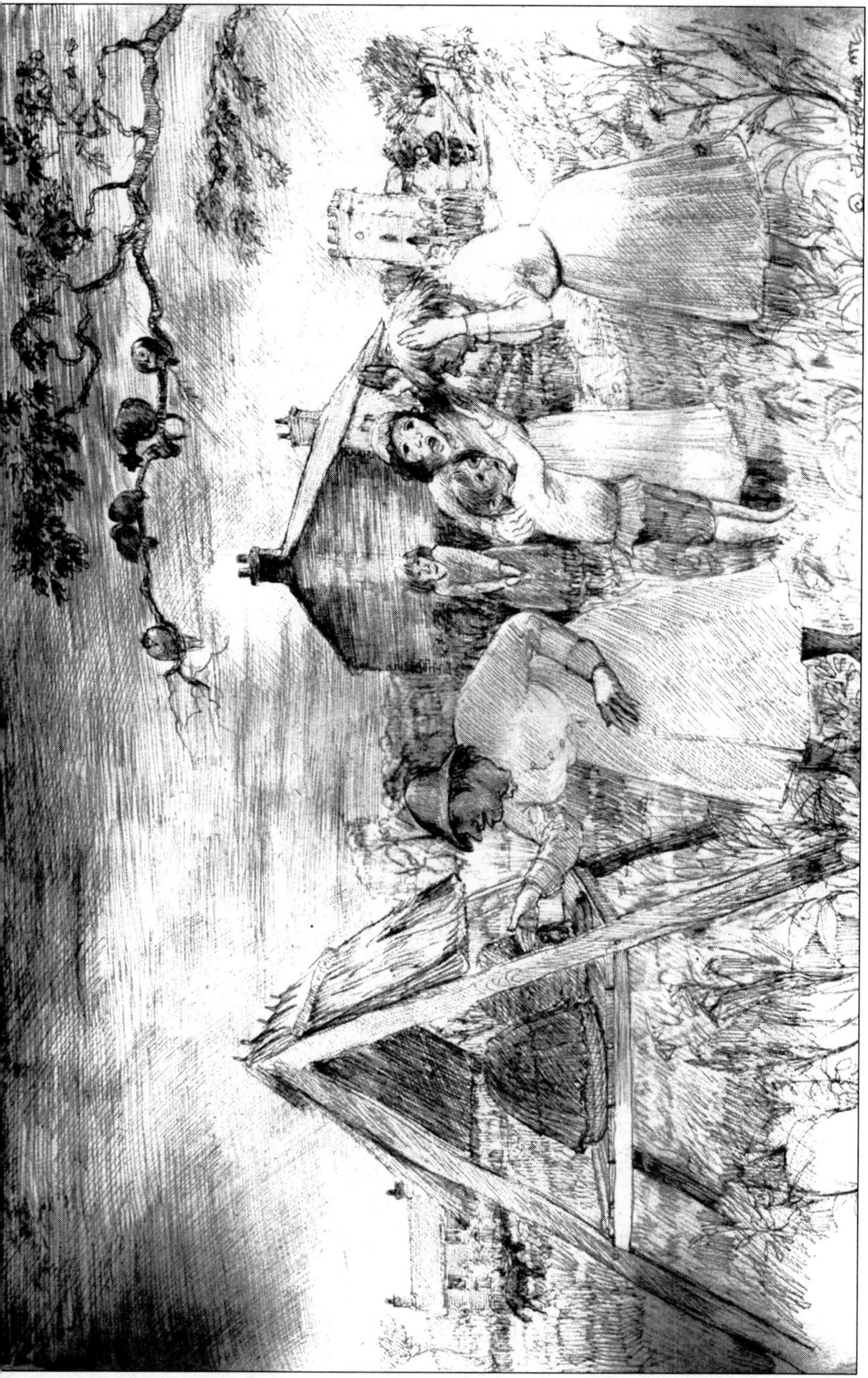

'Christmas Time'

Christmas approaches and the bees must have their cake of candy. Agnes has been given the privilege of placing it over the bees. It is cold enough for the bees to be safely clustered, though younger sister Ruth, looks on with apprehension.

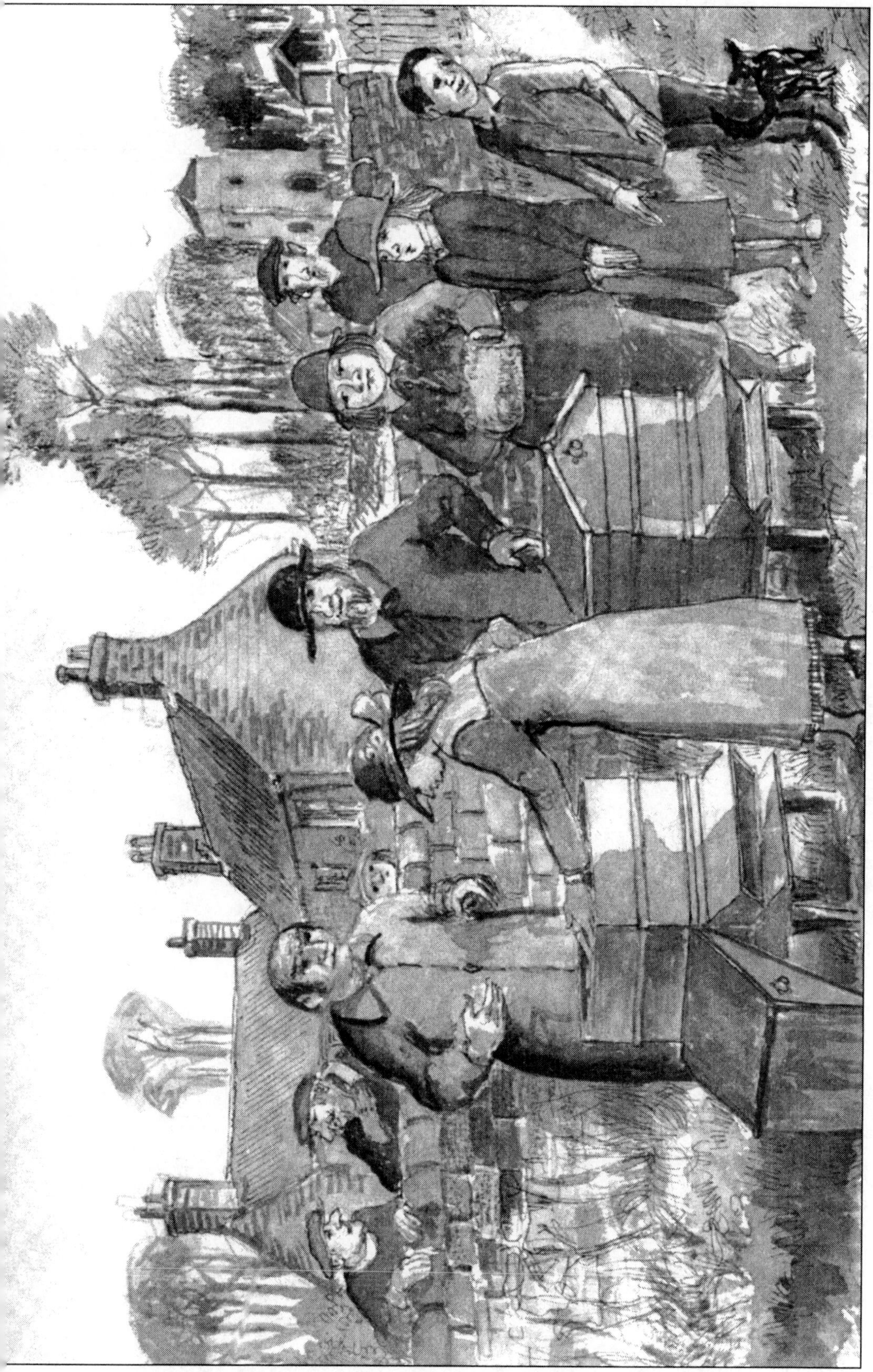

'Christmas Time'

Custom dictated that the bees were given candy at Christmas time. Father lifts the skep and Alice gingerly places the cake of candy on the floorboard beneath the cluster. Family and friends look on with a mixture of amusement and apprehension.

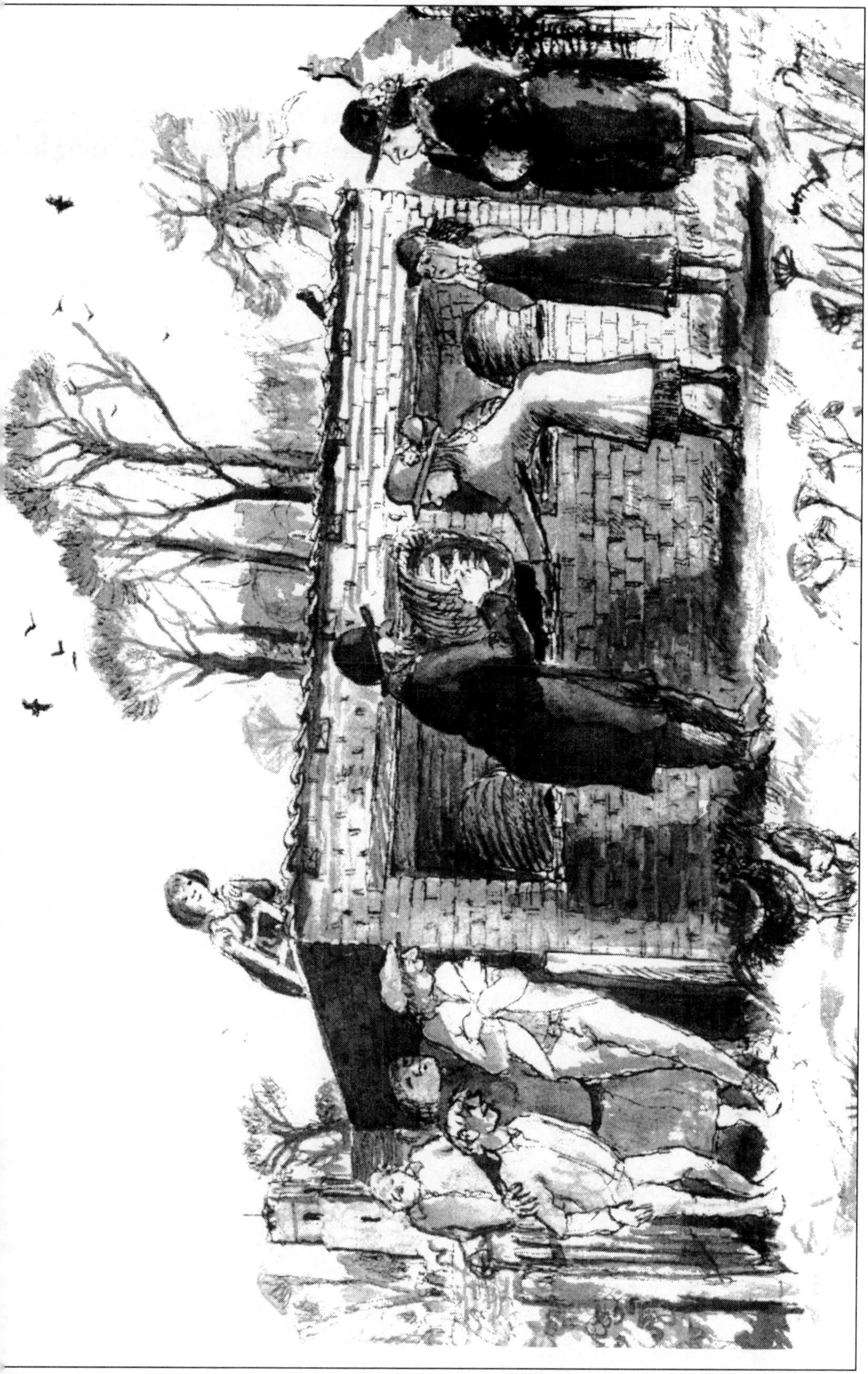

'Killing the Bees'

Father and daughter at work on a quiet cool evening in September. Gently does it before the bees have time to realize they are being led to a sulphurous end.

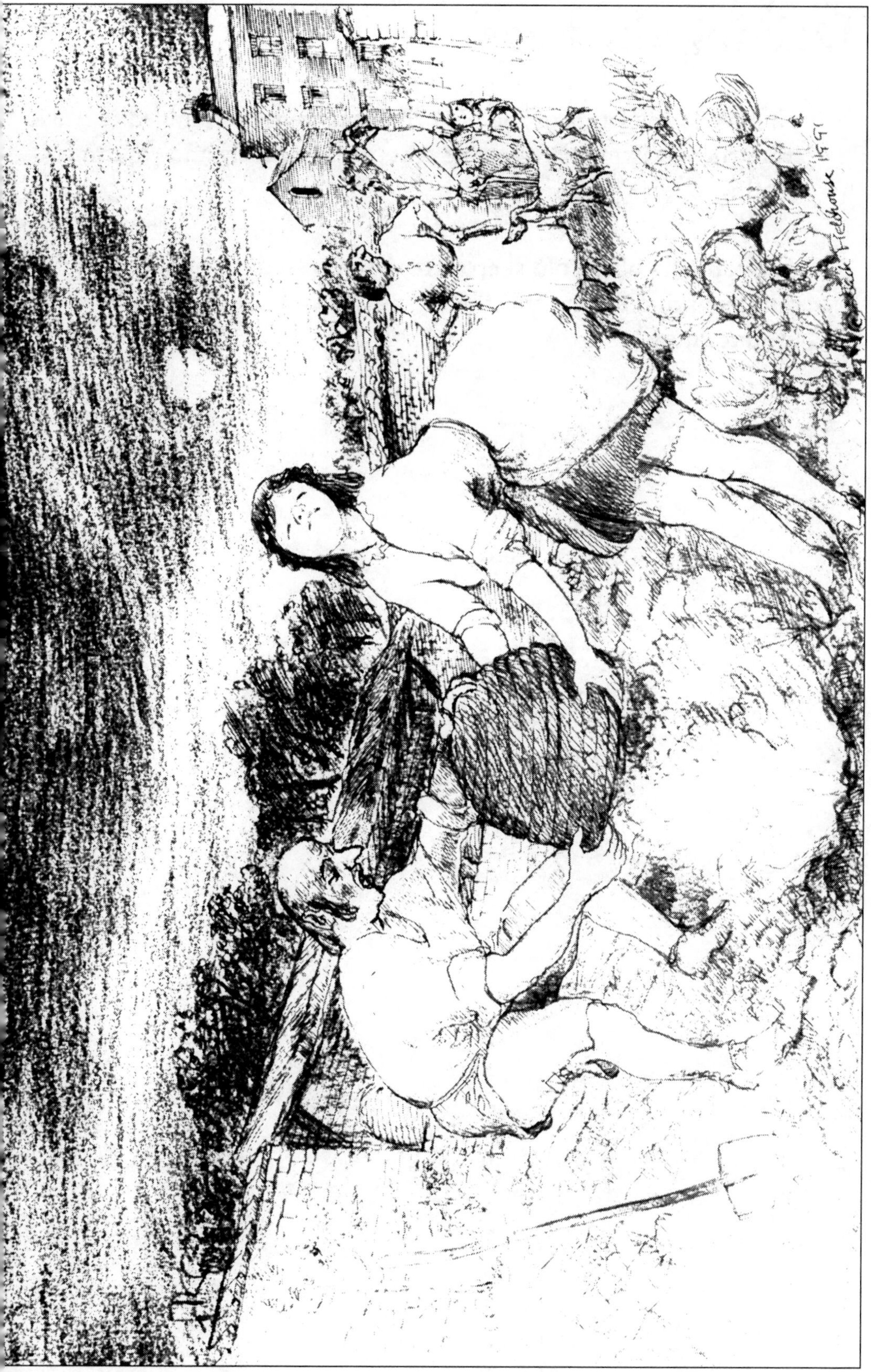

'Driving the bees'

To harvest the fruits of the hive, the bees were encouraged to move from their existing skep to a new home by a gently and rhythmic beating on the skep side.

The contents of the old skep were then pressed for honey, wax for candles and following this the remains were soaked in water to produce mead.

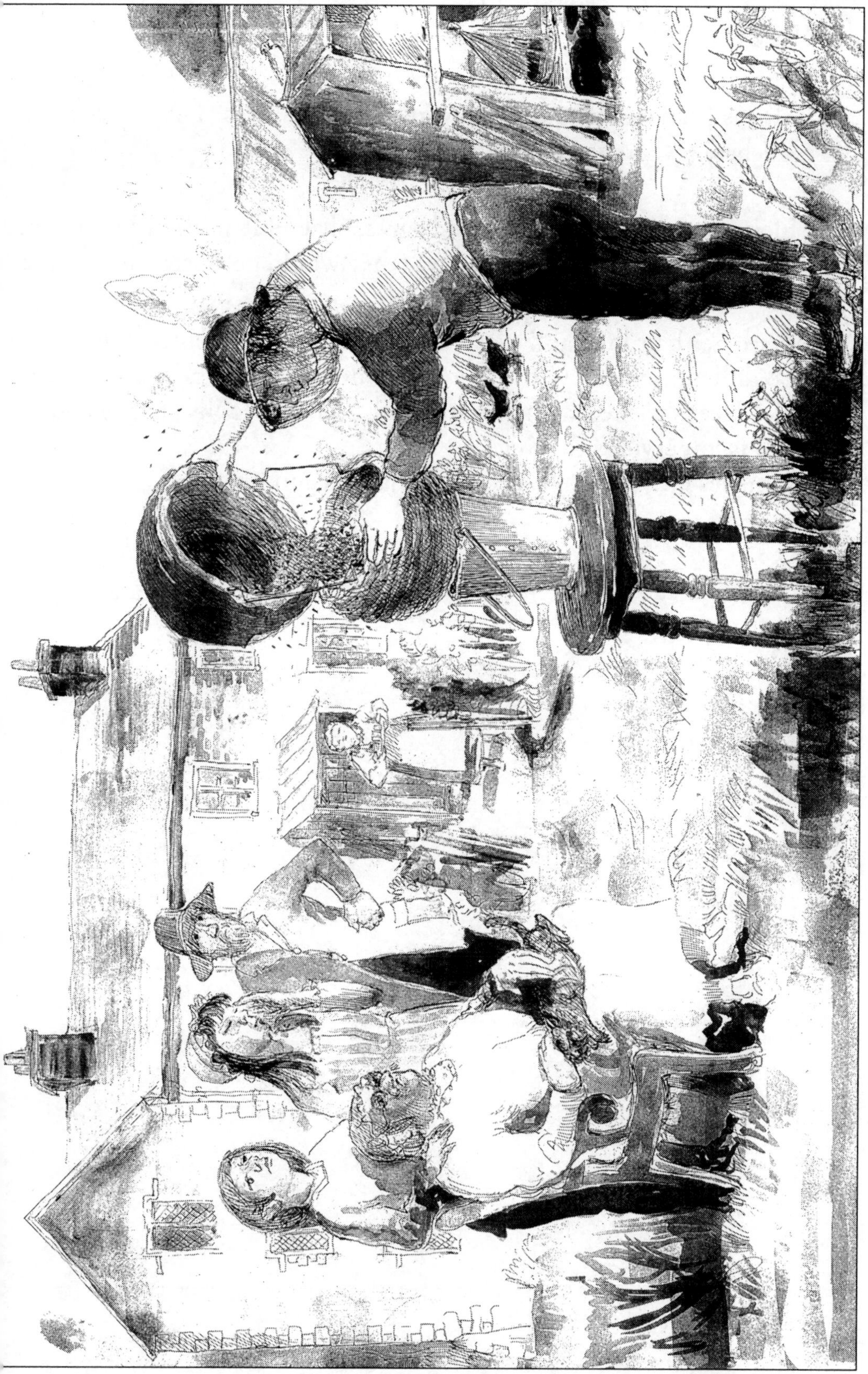

Aristocrat

This is Madame Bon Bouche with her three hives (Les Troise Ruches).

Her presence is aristocratic and agreeably emblematic of her manner and lineage, even the bees behave (they would be riotous to a knave)!

With her hair long and scented she avoids the demented state:
So difficult to deal with when taking honey and an added pacifier:
Her face is packed with poudre most French it looks ghastly and is
and is impervious to stings.
It makes harvest easy a nectar for queens and kings.

Jack Fieldhouse 201

Les Troise Ruches
Jack Fieldhouse 07

Les Trois Ruches.

J. Fletcher 57

On the way home from work

On the way home from work Reg Sendall always looked for a sign that Mrs Honniker needed help with the bees.

The sign was always a garment (bright) left on the hedge outside the house.

If there, Reg would stop his car and go in to see Mrs Honniker.

Mrs Honniker was eccentric and lived in an old cottage, the bees were kept in W.B.C. hives and she had an old one-legged table for supporting her combs.

"I allus takes off honey with a spoon" she said "I nivver nivver threatens them little dears with a knife!"

Bridgwater 1970

Mrs Honniker is quoted as saying (by Reg Sendall) "I allus takes off unn with a spoon, I nivver nivver threatens they little dears with a knife!"

Mrs Honnik

Mrs Honniker is quoted as saying (by Reg Sendall) "I allus takes off 'unney with a spoon : I nivver nivver threatens them little dears with a knife!"

Inheritance

Daughter julie thought sylvia was poorly

Went to see her as she was finishing her gruelly breakfast

The doctor came in:

T'is my urine she said

Julies mind was filled with dread

How would one know

Relief: it was an accent locale

Had brought confusion

The doc. Soon banished the illusion

'Your hearing' he said 'is not good'

'So soldier on my good mrs flood'

Now picture gran sylvia trying to hear

Through her idea of a bee proof gear

Imposs ible and vision nil

They decided to go in for something to fill

Their viewing need.

'Netting' she said 'and with my urine bad'

When e says summat ill ear

And be ever so glad

An eery tale

Whether youm ome spun or modran
A happy summer with the bees

WHETHER YOUM OME SPUN OR MODRAN
A HAPPY SUMMER WITH THE BEES
E.D.

Haughty Hortense

It's Hortenses' Day and I've completely forgotten why she chose this masochistic exercise in shaking combs onto a ramp and into a box on the original site with the sleaze of an excuse to frustrate the critters into changing their norm, which is to swarm.

Once tried never forgotten and for Hortense it's the first and last and her so called friends are aghast at the release of countless stingers, all hell bent on revenge. Such a quick retreat for those fleet of foot and panically inclined blame them not, they're sound of mind and norma folk are not inclined thus to stir unnecessarilee the hard working and precious honey bee

Jack Fieldhouse August 201

Hortense demonstrates A shook Swarm
Jack Fieldhouse 2007

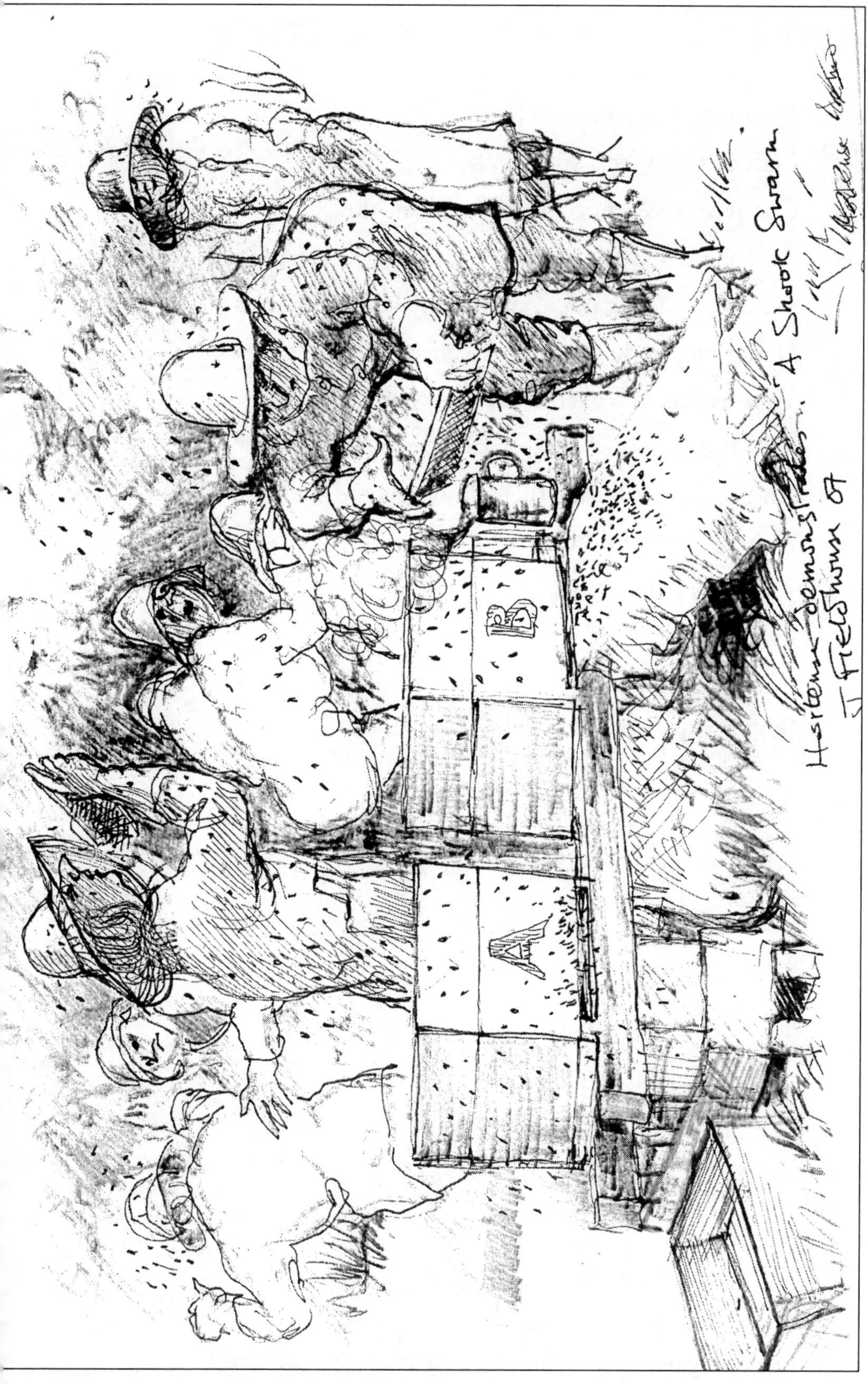
A Shook Swarm

Sympathy In The Vernacular

Maureen she got the itch

I telt er she were a silly bitch

Cos er veil werent tucked in praper like

And she got stung on the ed and what a mistoike

And worse she panicd, ripped the veil from er ed

Daft thing to do that to take off er 'at'

With er face uncovered the bees got really ot

And she got more stings which accounts for the pose

And er skin will turn the colour of a red, red rose

And be there for days

As i allus says it pays

To take more care

The silly mare!

Will learn

Jack fieldhou
July 201

The Stung one with onlookers only partially concerned

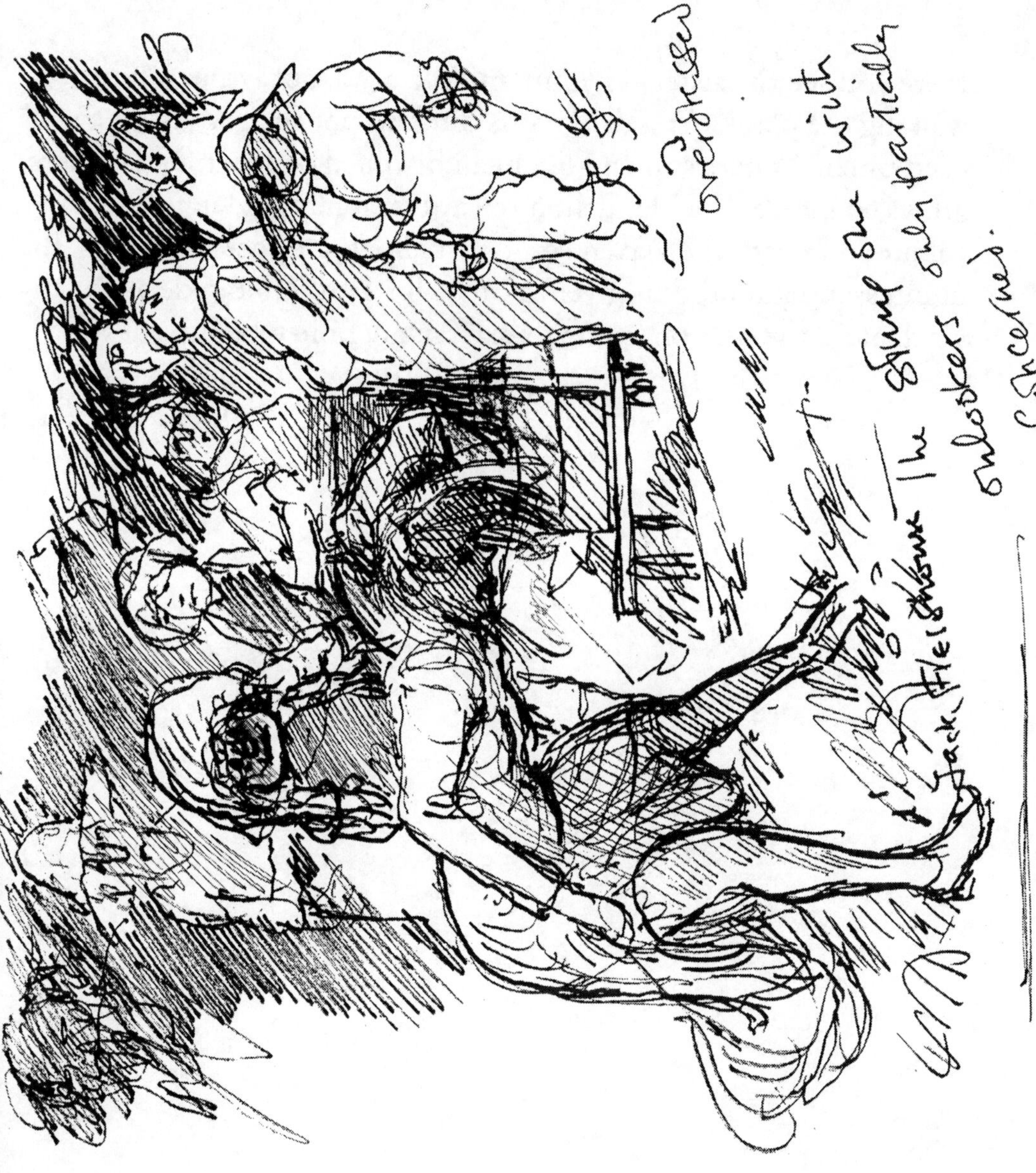

Grimm Reading

Every flower on those ten thousand acres of moor is temporarily theirs, it's a crop for those who dare to take bees to the 'weeds'.

Here you see the sisters Grimm, middle aged, but lythe of limb, walking the plank o'er a dangerous ditch to put their hives on a well used pitch. Where sun and light can be relied on, next fields where grows en masse The Dandelion reward, a honey golden yellow in colour and neath the waxen comb is mellow. Dee gee for weeds, the quintessential food thus given, a cause for sanity to rankle as to why we deplore the best oh, come to tea, try my honey, be my guest!

Jack Fieldhouse 1984

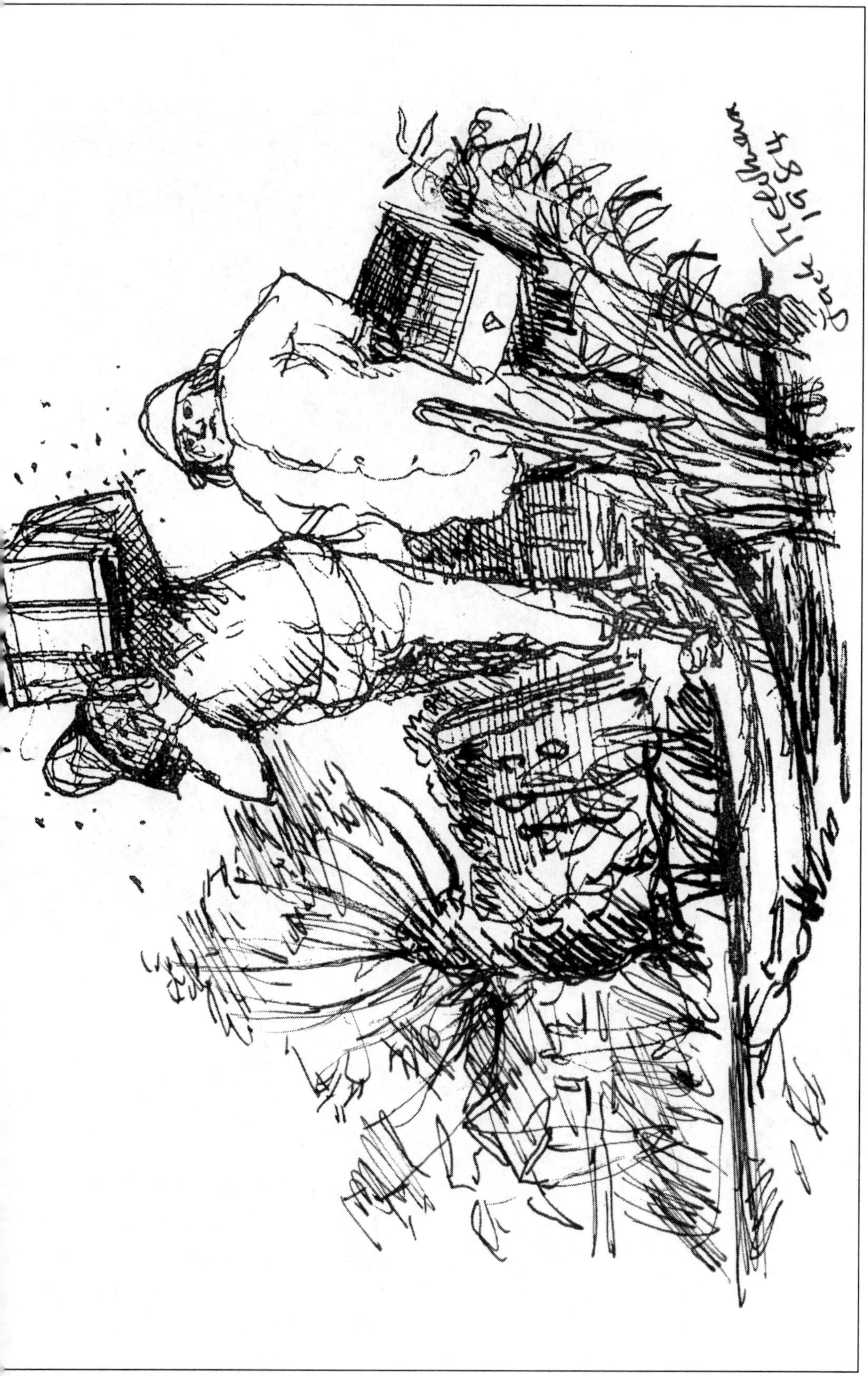
Jack Freedman
1984

A Mediaeval r
The two girls feel trappe
The Beekeeper and h
and the pet stork thr

Loo
autiously approach the swarm
The Owl is imperturbed
Jack Fieldhouse 07

Doc Smith's Petherton Plot

Flat, it is decidedly not, it's a lumpy garden of banks and steps and geese, a 'difficult to manage' piece and with space so short the Doc. ha had to resort to the vertical.

The bees don't mind but manipulators must find patience, especially getting down to lower layers and on each box, a covering cloth so imperfectly done so much shifting and shuffling and a mounting indignation from wrathy bees.

"Let's assemble the boxes where they were and get out of this stingy situation <u>please</u>!"

Jack Fieldhouse July 199

DR SMITHS APIARY N PETHERTON JULY 1992 Tracy Hedderwick

Cliff Apiary

The Bee Meetings organised by the Mansfields, Sid and Rosie was always well attended and ended with tea and cakes.

The moment deals with Rosie being stung in an unmentionable part of her anatomy.

Although Rosie attended lectures on beekeeping and the habits of bees, she wasn't paying attention when the lecturer told the audience that bees always climb up.

Cliff Apiary North Petherto
circa 197

A sudden stir,
Rosie Mansfield at the
Cliff Adam

Monsieur Bonhomme

Normandy and Varroa and Monsieur Bonhomme surveys his 18 hives!

What workmanship and before Varroa he could take up to a ton of honey in a good season.

Came the clever Biologists who thought that it would be a good idea to bring bees from the Far East and incorporate their qualities into our own bees.

Unfortunately they thought only of the benefits.

They brought in Varroa.

Beware of clever people!

On looking over this Apiary of 18 hives only one contained bees.

French Apiary

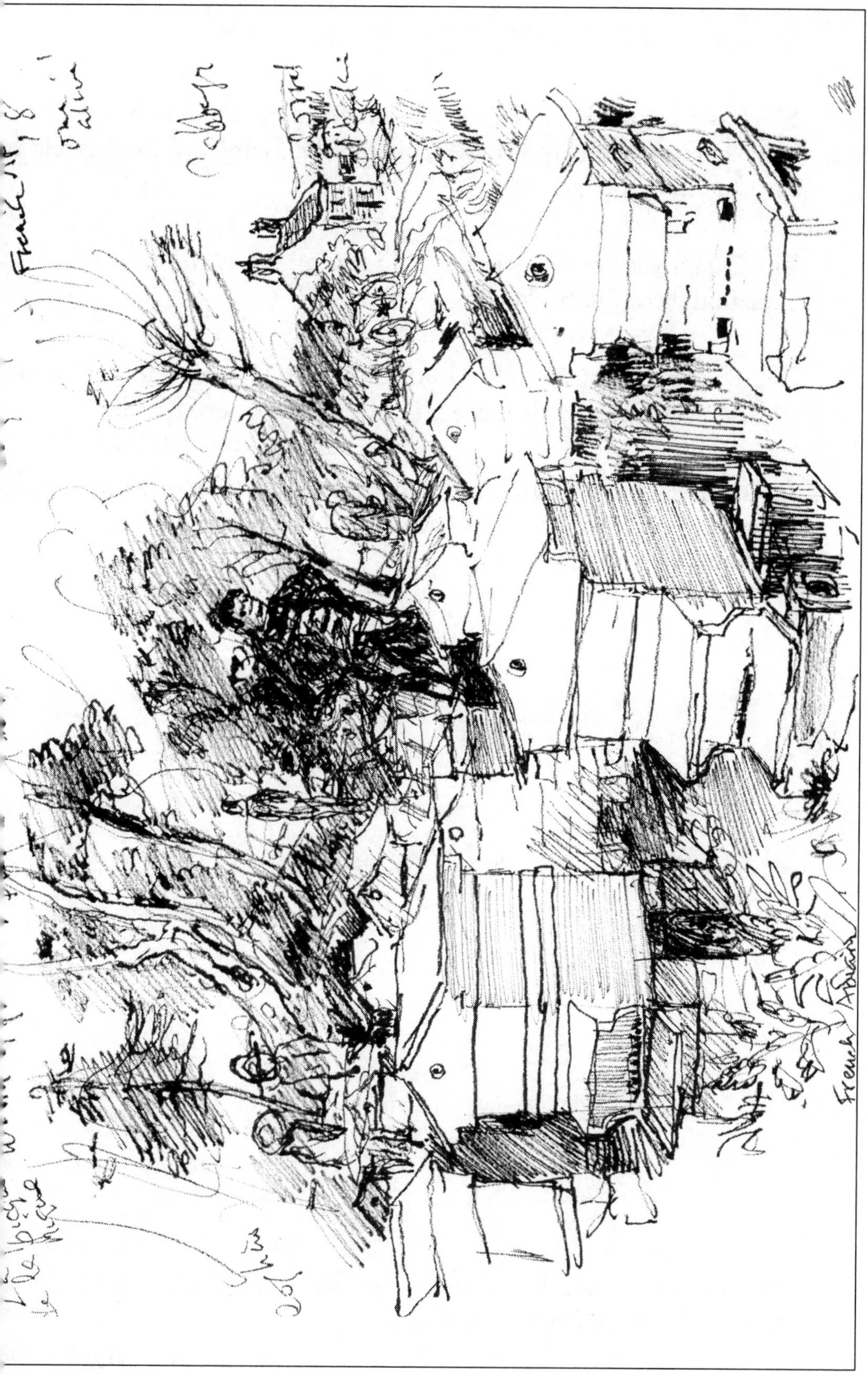

This is 'Summer' aptly named

She has a Teddy Bear, which she brought to the Quantock Beekeepers Open Day at Fyne Court in the Hamlet of Broomfield on the Quantock Hills.

The Quantock Bee Association hold this open day to the public to explain the craft of beekeeping.

There are demonstrations and side-shows and / or children under five if they bring a Teddy Bear there is a free small pot of honey.

Interestingly Fyne Court was the home of Andrew Cross, the pioneer electrician and it is ironic the Broomfield was about the last place to be connected to the National Grid!

Summer with her Teddy at Broomfield's Fyne Court to claim her honey from the Quantock Bee Association.

August 201

Summer with her Teddy at
Broomfield's Fyne Court
to claim her honey from the Quantock Bee
Association Aug. 2010

The Italians

Such an ordinary day, the bees placid and they stay calm under the gaze of Auntie Flo who claims her 'Italians you know are calmly disposed' and so much so that her 'naice' nieces can be invited to put on hats and veils and be delighted to be thought of as potential keepers.

'But no' says Sylvia 'If a bee wants to sting me on the knee it can, cos I'me not covered there and the other a Mary 'me same t'is alright for them, they covered all over. 'Then a shout from Aunt Flo 'There's honey coming in from the clover, the calm continues.

No flexing of sinews, placid reigns the mother hen with her chicks search for grains.

From Sylvia: "It's fee fi fo fum think I'll copy my aunt and stick out my bum!"

Hive inspect with ---------------- intruders
Jack Fieldhouse 2007

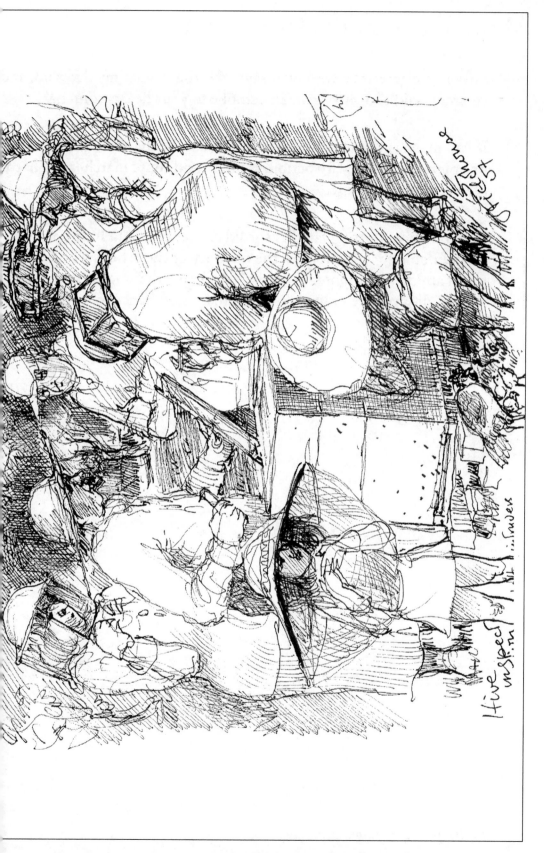

Hive inspection, Sinders

War Time Enlightenment

Fifty free cigarettes a week you say? Yes, that's what my dad said, 'and you may' said the Guvvernment, 'smoke to your hearts content.'

If you manage to find a moment to rest and be thankful and you're in one piece and what with action, training and route marches the hard life neer failed to cease.

Without a comforting fag this sordid existence would have been unendurable 'so lets smoke to the memory of those boys who bore so much amidst the frightening noise. And now the stage is set to ban the fag. It gives the politico/health fanatics a/chance to wag!

Smoke on, (sensibly).

Jack Fieldhouse Aug. 200

Beekeepers enjoying a quiet fag.

Beekeepers enjoying a quiet fag.

The Mansfield's

Sid and Rosie Mansfield ran the Bridgwater Bee Keepers Association for many years.

The drawing shows Sid as a boy accompanying the Treasurer of the Bee Club as it was in those days. The Treasurer had no social skills and was most abrupt in his dealings with members.

A woman obviously taken aback has to listen to "Now, be yew gonna pay Missus or baint ye, cos if yew baint I gonna scratch ee orf!"

'Now be yew Gonna pay Missus or Baint ye, Cos if you baint I gonna scratch ee orf'

An incident in the life of Sid Mansfield who accompanied his Bee Club Secretary to collect subs.

192

'Now be yee gurna pay Missus, or Baint ye? Cus if yew Baint ye gurna scratch ee off.' An incident in the life of Sid Mano fleet, who accompanied his Bee Club Secretary Castle of Subs 1920

The Oh So Clever Ones

"What wonderful characteristics" said the Bio Scientists – of the Far
East Bees. So let's bring some back and breed from these clever stuff
but unknown to them a parasitic mite (Varroa) came too the phlegm
of the narrow minded!

Once discovered it was too late, it swept through Europe at an
alarming rate.

A decimating scourge and at local level Monsieur Bonhomme says
"What the devil!" Only one hive left out of a healthy twentee and
what do I think of Bio Technee?

"UNPRINTABLE!"

Centuries old systems of renewal and harvest **gone!**

But in time Mother Nature (she will always condone). The pestilenc
will wane and Bonhomme's son will harvest honey again.

Jack Fieldhouse Aug. 20.

Width of frames --------- French Apiary - one hive

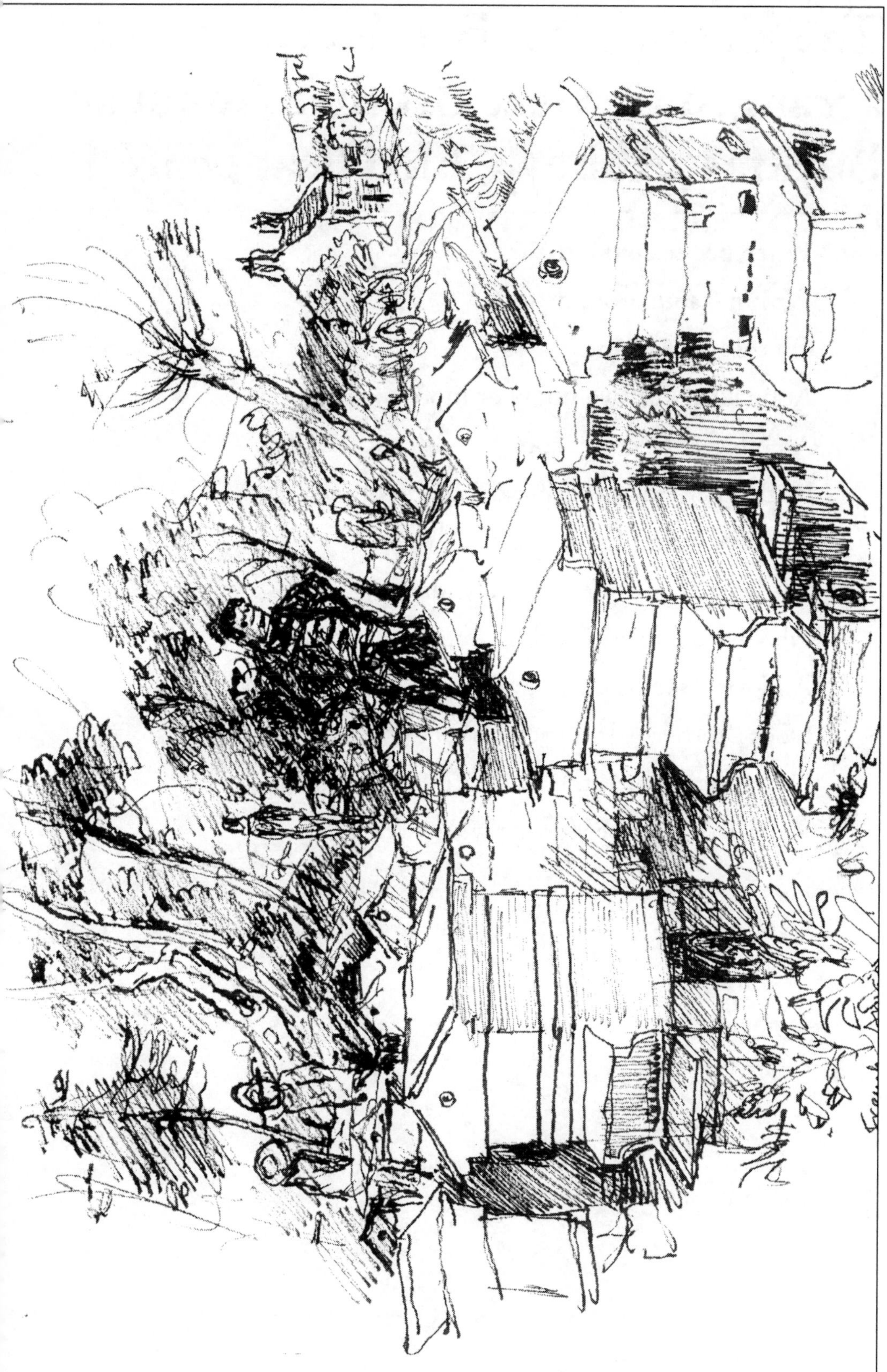

The Zummerset Bee-Keepers had arranged to Meet Brother Adam at Buckfast Abbey : He Disappeared

When bee ee conim then?

We bin stanin around

On this oily ground

At buckfast for two hours or more

An our feet be gettin sore

Its Brother Adam we're waitin for

Poor eve waitin for er adam

She kept er gear on – the silly madam

Oh what a frustrating day

And we come all that way

Does Brother Adam prefer bees to beekeepers?

Or has he reached the 'ah firgitts' stage?

A day to temember for the wrong reasons

1970

Bro Adam Forgot!

The Non Event, Buckfastlei

Bro Adam forgot!

The Non Event
Buckfastleigh

A Sacred Moment

Missa Solemnis?

Well almost: The High Priest knows there is just a ghost of a chance of finding the queen, why look? To make sure she's there. Nice to see her of course but the full glare of light will drive her to hide. Her instinct tells her not to abide such interference: So blend she must under courtiers numerous.

So that those with their lust for searching will abandon and frustratingly lurch toward tea and talk and noise and adopt a casual arm bending poise, meanwhile the queen now returned to her gloom and her egg laying work she at last can resume.

Jack Fieldhouse Aug. 201

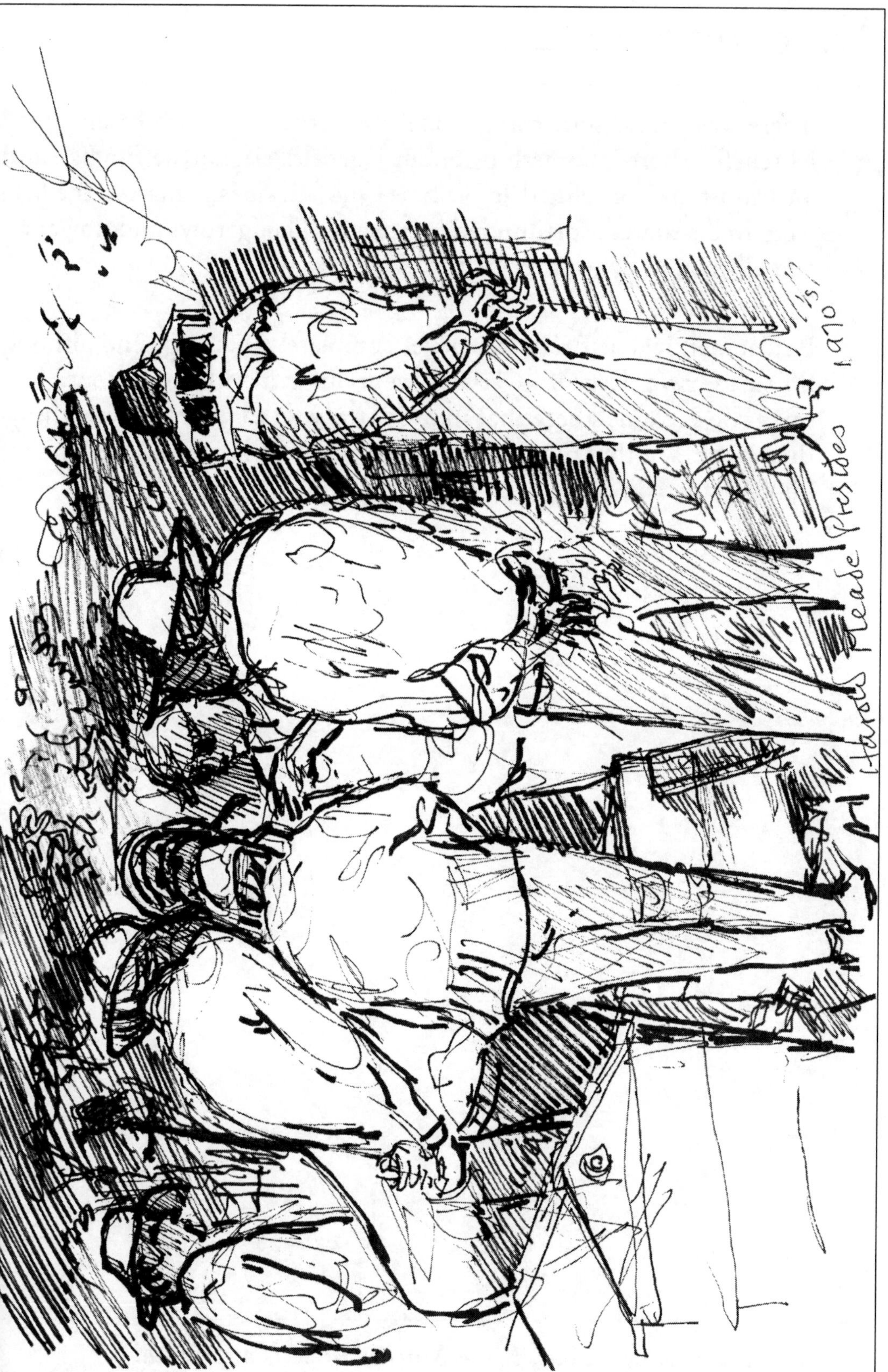

That Smoker

There was always something casual about the meet with Brian Mitchell. There is something about Fairfield, elegant, so English and an atmosphere of refined 'Je Ne Sais Quoi'. It doesn't matter, the bees were well situated, surrounded by trees and hedgerows and flowers and all created by Mother Nature.

But, somehow our Brian was never completely in charge and although his smoker was large it invariably went out at a critical moment. It might be explained here that in order to avoid foment the only thing that matters at that moment is smoke and if its not there, well, hubbl bubble, you're in trouble!

Jack Fieldhouse 200

'B….. Smoker's gone out'. Brian Mitchell inspects at Fairfield

'B..... Smokers gone
out.' Brian Mitchell
inspects at Fairfield
Stogursey.
1984

Jack Fieldhouse
'87

Suddenly

Suddenly the bees erupt

It's the thing they do because they are apt

To respond to a careless move

On the part of a keeper trying to prove

He can cope, when he cant

Comb dropping or a sudden jerk

And they don't approve

So they dive bomb the heads of all around

And woe betide anyone found

To be lacking in armour

Cos if theres a chink

They are in, and the worst scenario

- There could be a stink

Amateurs beware

Approach with care

Calm produces balm

The opposite is panic

Boxes everywhere

And a sudden rush for safety

Yes they're wild!

Bad Tempered Bees

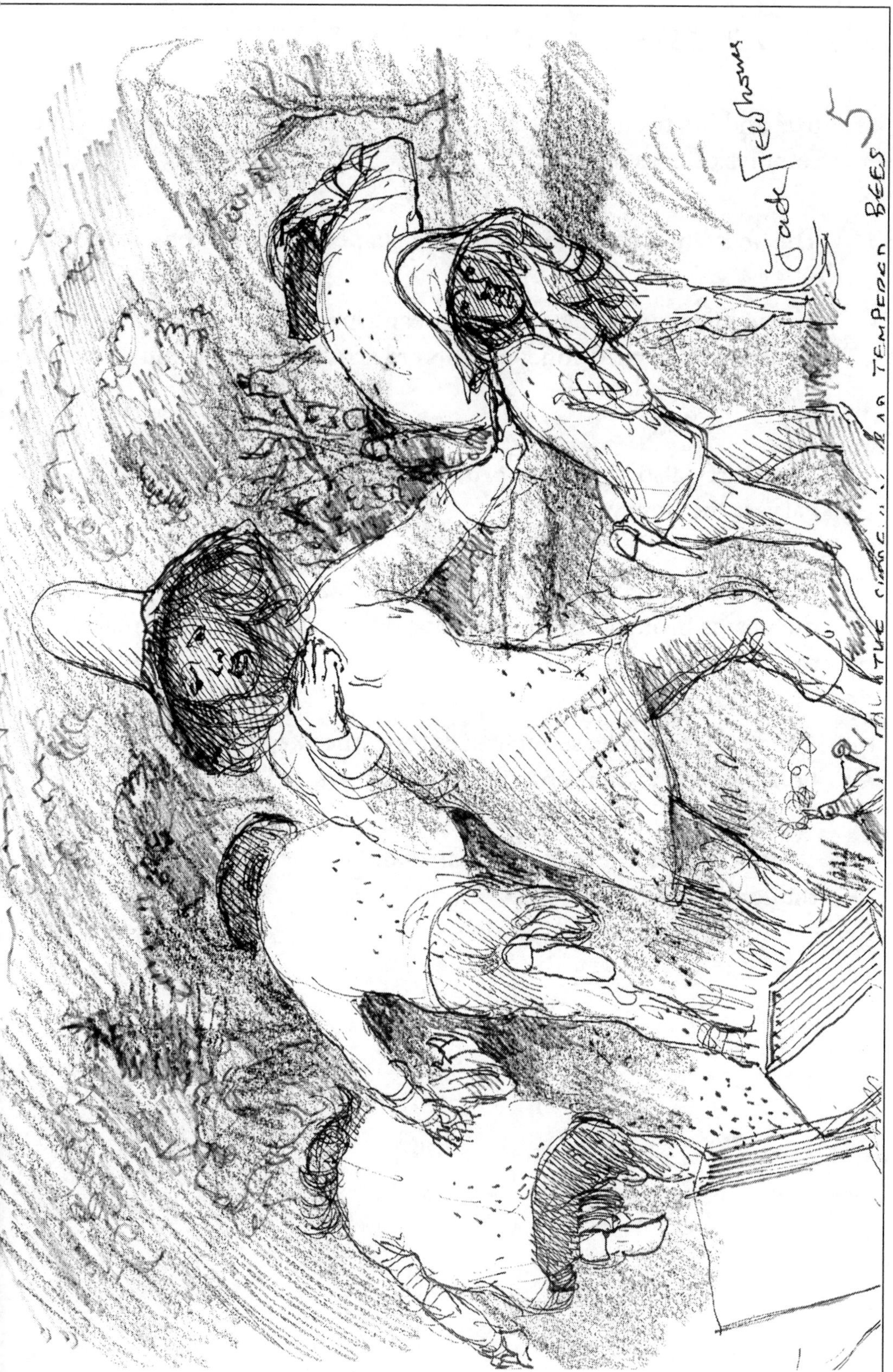

The Spanish Honey Gatherers

José and his friend arrive at the caves where they know there will be colonies of bees.

The bees build their homes of wax suspended from the cave roof.

In late July when the honey has been ripened and sealed with wax it i the time when the Honey Gatherers arrive with their mules each with a skep of bees on each.

They carry also a gun and provisions for the few days it will take to harvest the honey.

Within the cave, the nests are located and with a shot from the gun the nest is broken, the honey drips or the waxen combs break and fal to the floor.

The Gatherers release their bees from the skeps and they gather the honey which is then brought to their own.

Within a week the honey is sealed and they can return home with the 'loot'!

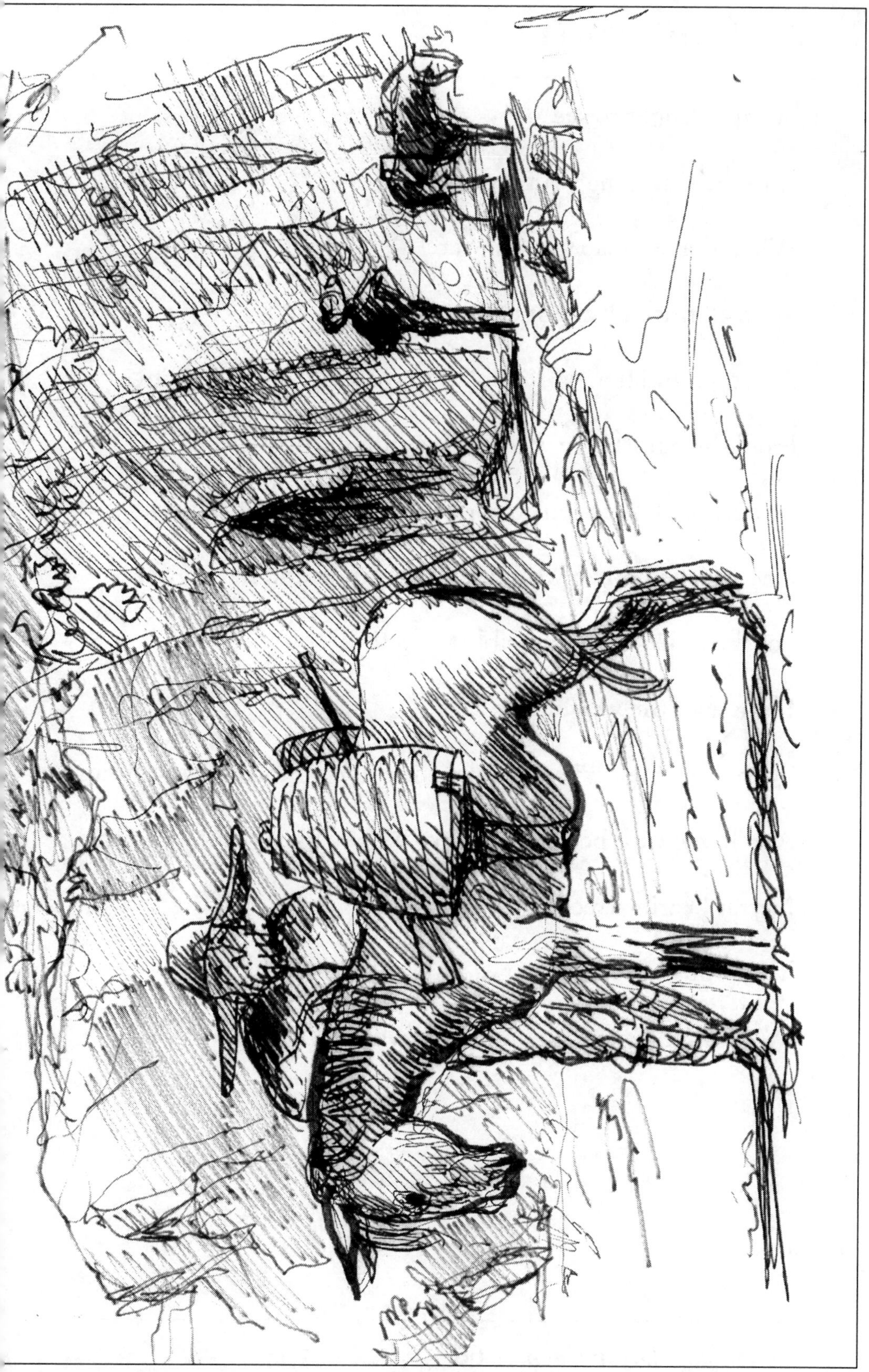

Bees And India

What a delight to see

And she's bringing a tray with tea

When the bee inspection's over

With all this talk about clover

And rape and honey from blackberry makes such a change

From the Sainsberry when hither and thither with basket

And list and directions from her and my mind in a mist

And weighed down with fizz, cheesy milk and more grist

For the mill: she's impossible to find!

I lean with the weight

Soon over: the till in sight and the stat e of normality resumes.

And homeward bound i think of bees of honey and wax and

Nectar bearing trees

The bonus of course, comes surely at four

With sammidges, cakes and pots which pour tea.

Bliss!

So welcome Jack Fieldhouse 06

So welcome!

Jack Field Jackson
Jack 06

A Cornelius Inspection

The advantage of being a beekeeper in the Middle Ages was that there were no bee meetings and interference by bee keepers.

Theft was a problem and here you see Cornelius and his wife inspecting their skeps of bees.

Cornelius a strong man made platforms in a tree for safety.

In August the skeps were tested for weight and the heaviest and the lightest were taken to the sulphur pit and the bees destroyed.

The wax and honey would then be cut out of the skeps and served through muslin.

The wax was for candles.

There is a boy with his pet pea hen and the chickens lay their eggs in the dog kennel.

J Fieldman 201

The Flemish Scene

Cornelius and his wife check on the hives in early August when they are about to decide which to take to the sulphur pit to kill the bees and harvest the comb and honey.

The skeps will have the wax combs removed and put into a large dish where the whole will be pulverised and eventually strained.

Honey for the house and wax for candles.

Some skeps will be kept till next year when hopefully they will swarm and provide more colonies.

Flemish scene – Cornelius and his wife check on the skeps of bees
August 1601

Flemish Scene;
Cornelius and his wife
check on the skeps of bees.
August 2 1601

61

In Praise Of Mother Nature   1

It's April and it's Earthy Day and a week before the month of May when movement starts in root and bud and druddy folk give praise to God. Their God, that mighty magnet the sun which daily demands that the race is run to meet the lengthening days with toil without which the harvests of the soil would never be.

The scene is set: There's Fringy the Thinker the village clerk who frowns on Boumbig the Stinker, the hunter, begetter and prone to wonter with 'Kaaren' his most recently and deeply flare and well begotten kinder who come to stare.

But the now: A toast to her who gives unstinting the most in mead. And praise to the bees whose honey'd home kept by oer stung woorn provides the comb. Cheers!

Druddy – Druids

Wonter – Wantonness

J Fieldhouse July 201

---

in the Bee Garden, Fringy + Boumbig drink mead to the health of Moth Nature on Earth Day 22nd April

J Fieldhouse 08

In the Bee Garden
Fringy + Boumbis
Frank Mead is
the health of Mother
Nature on EARTH DAY 22ᵈ April

Bridgwater Bridge Scenario

From irate ma "you inconsiderate lout you're responsible – she's up the spout"! Loud of course for all to hear. He answers not: he is to blame whatever altercation ahead the nurse will say "welcome to the world" and let's pray. That togetherness will happen: Three is better than two, but ahead one can see a difficult mother-in-law phase. Wil she mellow and give praise to God for such a treasure? Condemn in haste repent at leisure. Blame nature but don't be contrite it will only harm the little mite.

Zey

MJ'Ackuse *Confrontation on the Bridge, Bridgwater*

J Fieldhouse 198

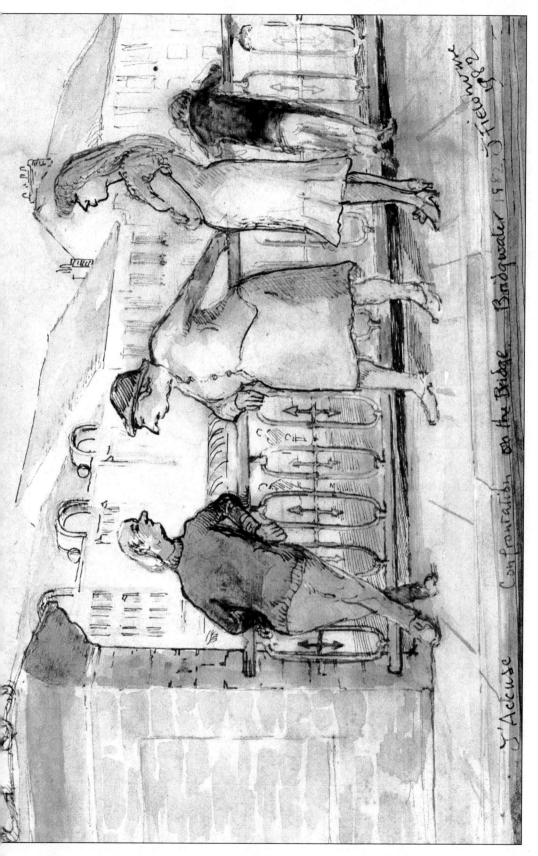

J'Accuse   Confrontation on the Bridge.   Bridgwater 1872.   J Fieldhouse

A Caravan Week in June 09 (near Tenby)

Tincture of Tomatoes
And pristinatious potatoes
Alliteratively stored in smelly Loo
Cos space demands
Agin the rules of 'GOV'
(Designed for effusive love)
of health and naivety to ensure
A sticky, smelly, social glue

And for the student of Art
To willingly part with scalpels
A week of Anatummy is recommended within
But be quick with your pad
To take advantage of the mad
Fleeting moment
When all or most is revealed
A week in caravanserai aroma
Is enough to help you with a good diploma

J Fieldhouse (

A Caravan Week in June 09 (near Tenby)

Tincture of Tomatoes
And pristinatious potatoes
Alliteratively stored in smelly Loo
'Cos space demands
Agin the rules of 'GOV'
(Designed for effusive love)
of health and naivety to ensure
A sticky smelly Social
glue

And for the Student
of Art
To willingly part with
scalples
A week of Anatummy is
recommended within
But be quick with your pad
To Take advantage of the mad
Fleeting moment
when all or most is revealed
A week in Caravanseri aroma
is enough to help you with a good
diploma

Jack Fieldhouse 09

Roche Abbey 1928

The Fieldhouse family often visited Roche Abbey near Maltby in Yorkshire. We took tea in the Gatehouse and if mother needed the toilet there was one at the bottom of the garden.

On one occasion Florence my mother met a miner emerging from the loo and to be told "I kept seat waarm fe you luvvie".

It was a smelly, disgusting and almost Mediaeel shed and mother if she could, avoided it.

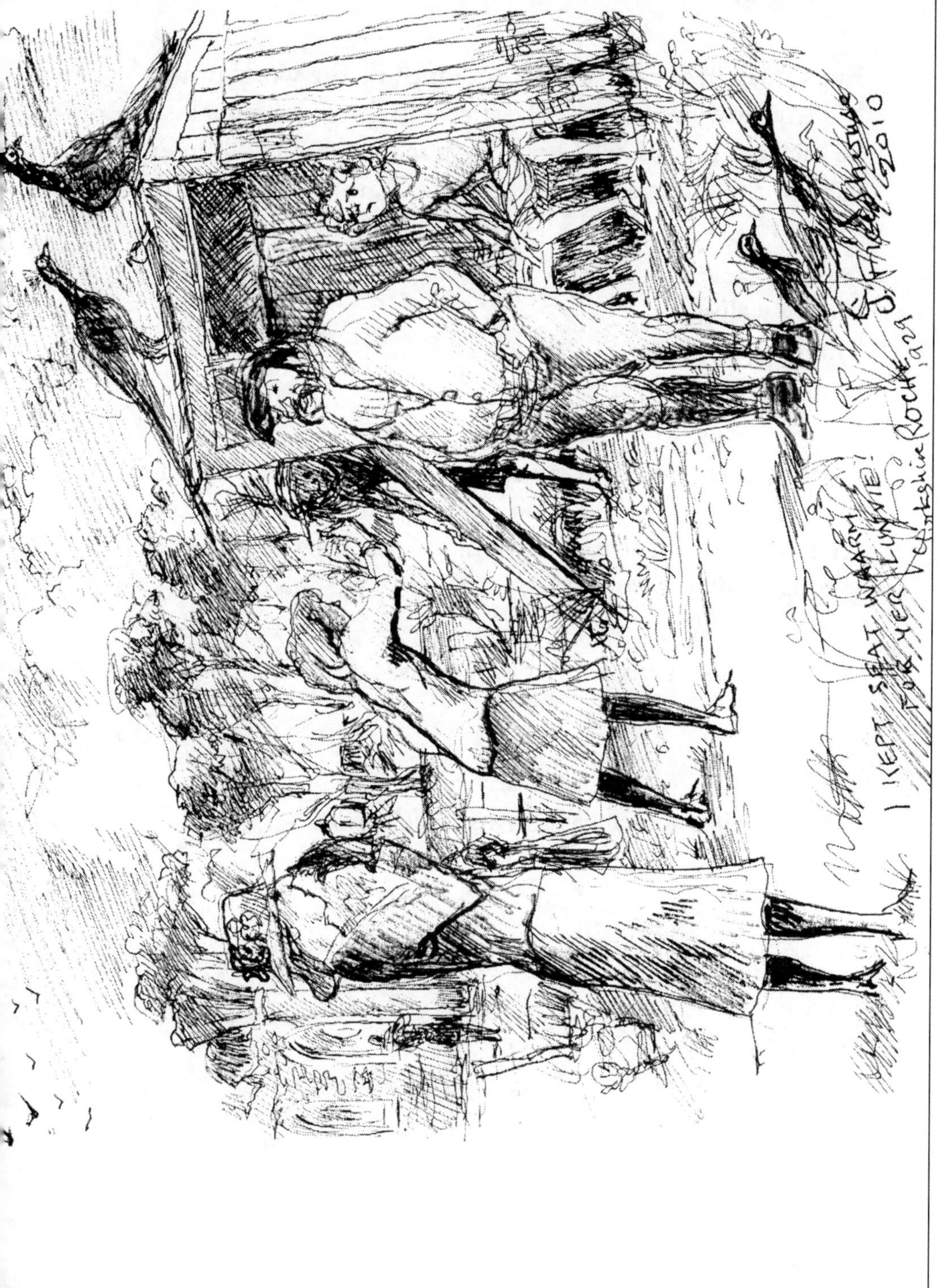
I KEPT SEAT WAARM FOR YER LUMMIE!
2010

Report From Ork

The brimming bath gives Burge a clue such blockages are nothing new.

But somewhere twixt hoose and outflow the work awaits – giving healthy glow to workers all, especially Burge who wields the rods wit all the urge he can, while others gaze in a glowing moment of a July haze so the question is: To goddess drain – where is it that you feel the pain?

Here a welling
There a swelling
A suppurating surface

A soapy puss, where normally it's dry as duss and not suitable for the likes of us say the dabbing ducks who make a fuss and quit the scene with bubbly gapes and 'yukking' as they go. The bath is draine (though a little stained).

The final gush brings forth a cheer at last it's time for an Orkney bee

J Fieldhou
Broomfield Hall (
And dedicated to my friends Mike and Sheila Burger, who live on Orkn

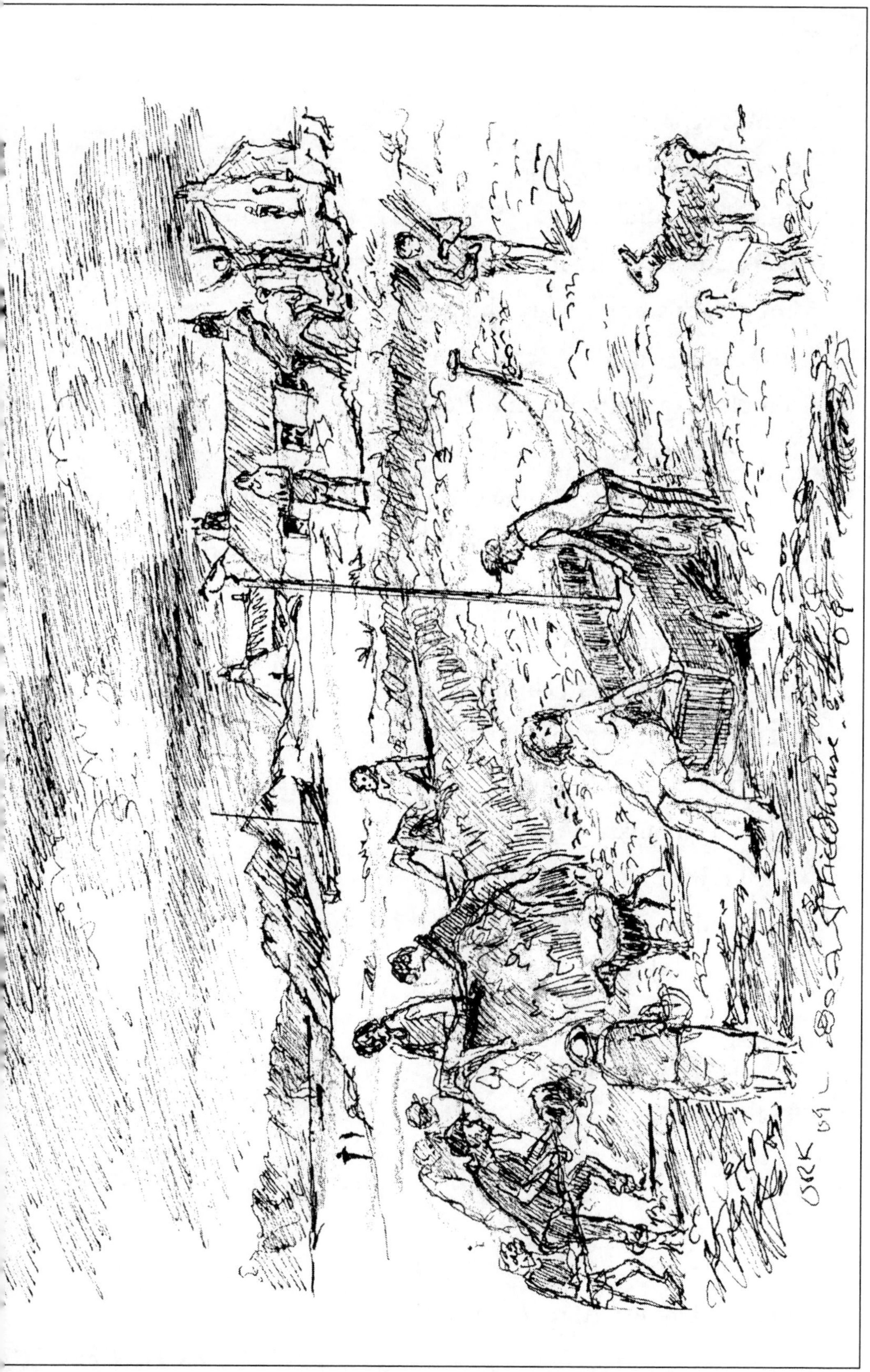

Ork

Windswept Isle

Grass and rock

First glance dull

Look closer : an artist has been at work

Grey black stones, lighter, higher

Exquisitely tinted by natures palette

Simple shapes almost childlike

Caused by lichens

Cadmium yellows, shining brick reds and greens

And all on ochreous layers of lichens past

Weather : windy of course

But on the tenth of February

Came the gale of gales

The henhouse of O'Burge was tipped into a field

And an oil tanker went for a dip

Such is life on a treeless isle

'Tho willows grow tight against the stone

(on the south side)

A perilous existence

So the O'Burge family

I salute you!

J Fieldhouse February 201

10.2.11, Orkney, Apres La Nuit Orrible

Jack Fieldhouse 201

"Après la Nuit Orrible" 2011
Jack Fedshoma

73

The Private Viewing

The Hoi Paloi have found a chair from which O'Burge can sit and stare. He told of his plans for all to hear of how he concocted this special gear. To hoist a something big and just in case it was bigger than big, to top his grace in story telling. (The fisherman's right to or and on go). No need this time to exaggeright! Cos here it is for all to stare at.

The niceties of how he began to beget this flaming flounder of unseeming proportions a friend who said throw your cautions to the wind and with hieroglyphic marker to boot and a hook abnormal cased in worms – astute to say the least, all held in neatly bag and if it bites be sure to lift it clear of wreck.

Just let it fight a goodly hour till sinews flag. When at that stage you might just possibly have it in the bag!

The dazed O'Burge could not believe his luck at hauling back this crook of the deep.

What is it? – A hybrid all agree caught in the wanton waters of the sea.

Hybridean of course and at last a change from Campbell bits on toast. Its chips now and flounder, which will we like the most?

J Fieldhouse 201

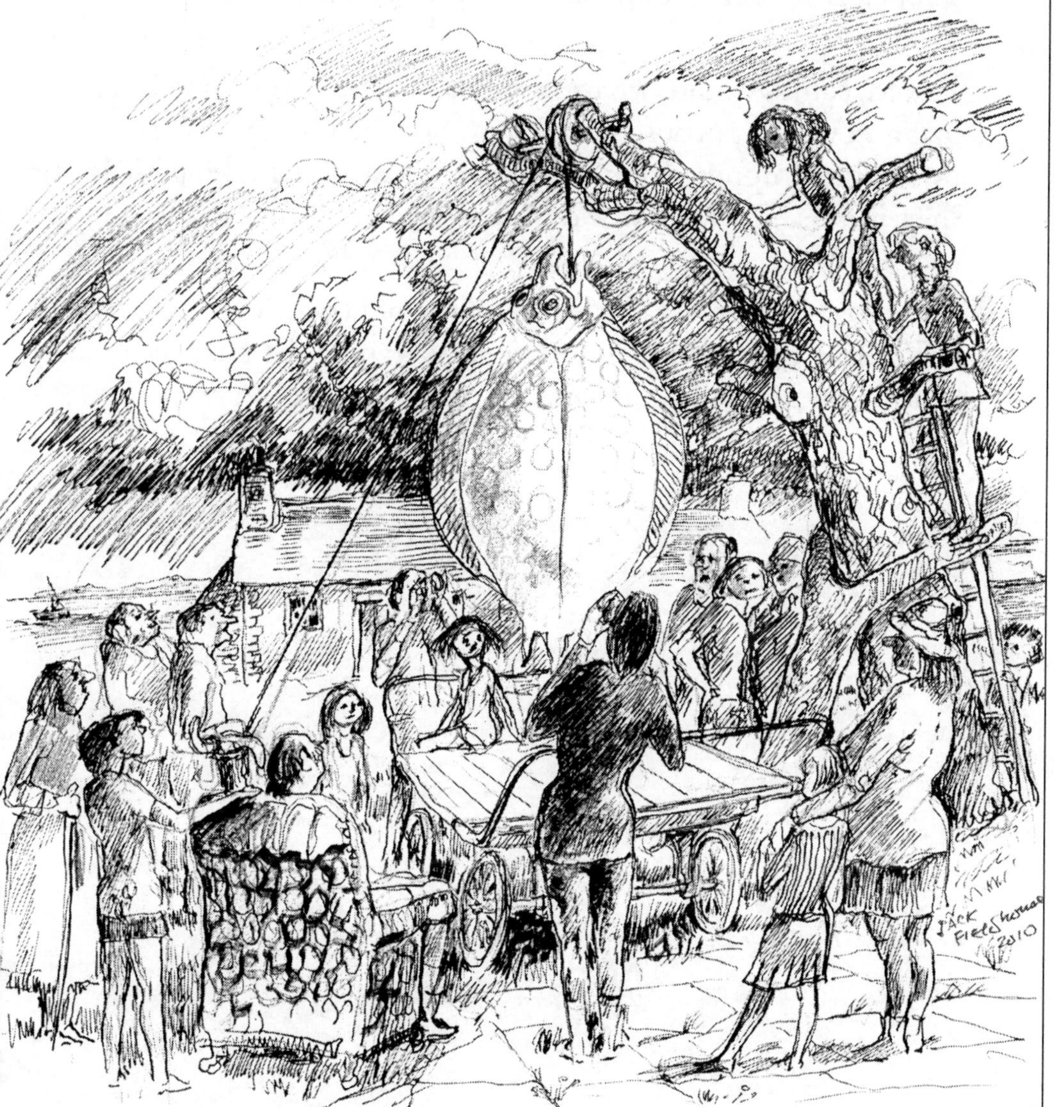
JACK Fieldhouse 2010

Thoughts on the launching of the Burger Boat in May 09

Lady Mac said it should

Skipper Doo said it could and O"Burge said it would.

Should, could, would, what? Built to float, it is of course Burgers boat and with a purpose which might be seen by a porpoise momentarily upping from beneath or those breathtaking birds the Shearwaters, in a wave world of their own, seen and lost in a moment

Whatever, O'Burge will fish and fill the old traditional dish the scrap to fill the hearthese cat. The over stroked mog the pusillanimous creature on the traditiional mat.

J Fieldhouse 0

The launch of boat O'Burge, Orkney May 09

J Fieldhou

The Launch of boat O'Burg. Orkney May 09

Burgers Embarrassment

The 'Ork' Des. Res. Lav. has a double 'U' handle phonetically musical yes but the inoffensive and humble 'thing' broke off, rendering the cistern armless.

Burge thought: Might the Cistercian chapel have a spare?

A gormless idea and the iron man said such fittings are dead and the islands gone metric and plastic.

But there's just one chance if you're prepared to glance through the island's dumps to find an oldie long forgotten neath nettles concealin' all that's rotten.

Meantime to borrow from Lady McDoo her ancient boat-ended crummy old loo to be set down 'mongst' Sheila's variegated netting gainst duck bills and sheep, their mouths all a 'fretting' and let's hope that Mike's searching won't be too far in resurrecting a dead and rusting old bar.

J Fieldhouse Sept (

And dedicated to my dear friends Mike and Sheila Burger on the island of Orkney

The Mobiloo in place temporarily courtesy of Lady McDoo
J Fieldhouse Aug 09

The Mobile-in-place (Temporarily) Courtesy of Lady McDoo

Jack Fieldhouse Aug. 09.

DEDICATED TO MY FRIE
ON THE ISLAND OF OR

ouse 09

E AND SHEILA BURGER

From The Kinders Ork:

We have a boat-shaped trolley
It's pulled by fat, sweet smelling Molly
And when there's a gale
We put up the sail
It's the undie we borrit
From Lady McFudd
And on it we paintit a chryssimus pud
And now we cart fish
For skipper McDod
And sell them to heinies
At the hamlet the noo
And they come with their dishes
To say 'oh my my'
And then buy a fish to take home to fry
And Mary the flounce
She does the accounts
And pimply Maddock
He sells them the haddock
The 'Dod' isn't mod
He pays us with cod
Or any we choose
To take back to the hoos.

J Fieldhouse 200(

Dedicated to my friends Mike and Sheila Burger on the Island of Orkney.

Anthea photos the reluctant gnome
J Fieldhouse 2009

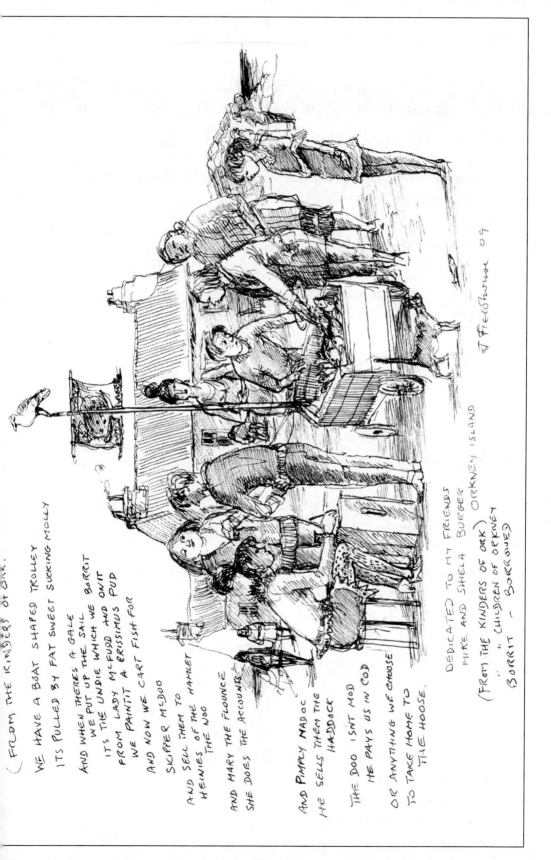

(FROM THE KINDERS OF ORK.

WE HAVE A BOAT SHAPED TROLLEY

IT'S PULLED BY FAT SWEET SUCKING MOLLY

AND WHEN THERES A GALE
WE PUT UP THE SAIL
IT'S THE UNDIE WHICH WE BORRIT
FROM LADY McFUDD AND OWIT
WE PAINTIT A ERISSIMUS PUD

AND NOW WE CART FISH FOR
SKIPPER McDOO
AND SELL THEM TO
HEINIES OF THE HAMLET
THE NOO

AND MARY THE FLOUNCE
SHE DOES THE ACCOUNTS

AND PIMPLY MADOC
HE SELLS THEM THE HADDOCK

THE DOO ISN'T MOD
HE PAYS US IN COD

OR ANYTHING WE CHOOSE
TO TAKE HOME TO
THE HOOSE.

DEDICATED TO MY FRIENDS
MIKE AND SHELA BURGER

(FROM THE KINDERS OF ORK) ORKNEY ISLAND
" CHILDREN OF ORKNEY

BORRIT ~ BORROWED

FredShaw 09

Port Mulgave (Yorkshire)

An area of Lias, well above the high tide but below the cliffs and only accessible to the fishermen or those athletic enough to cope with a steep climb.

From this Lias bed the fishermen launched their cobbles to fish for crabs and lobsters.

The sheds were made entirely from driftwood.

The man depicted, lived for five years in one of these sheds, living on winkles and fish and it is said that he never cooked.

When drunk he was ghastly

When sober most helpful

We're told he entertained the local rats!

One fisherman I met entertained his five grandchildren there for the whole of the summer holidays.

They made fires, cooked or barbecued, snared rabbits and at low tide collected mollures. Or they were crabbing with granddad.
What a wonderful beginning!

Cyril Cruakes five years in shed lived on winkles and fish and
he never cooked.
Ghastly when drunk
Most helpful when sober
Entertained the local rats!

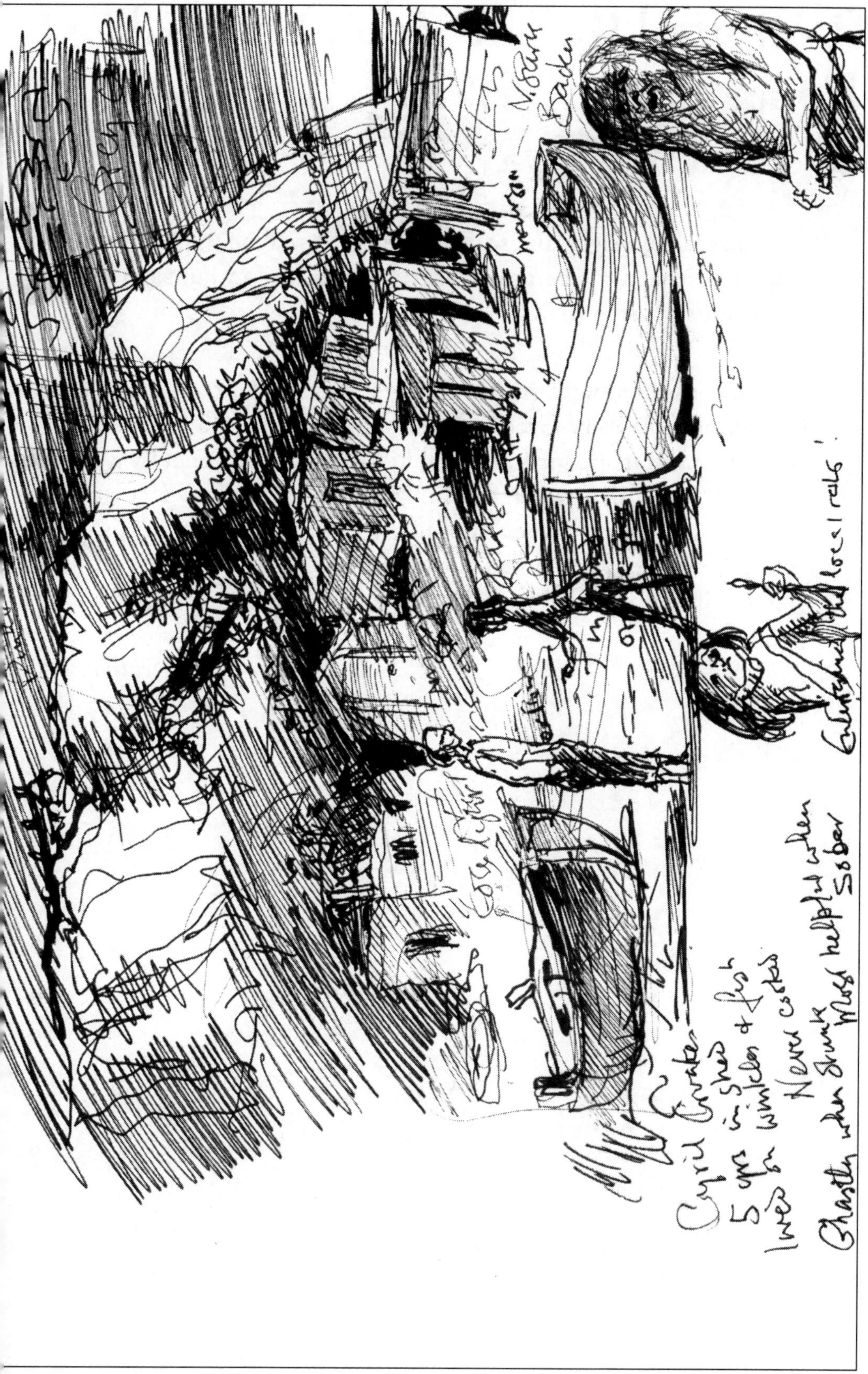
5 yrs in shed
lived on winkles + fish
Never cooked
Ghastly when drunk
Most helpful when Sober
the local rats!

Prompted By The Book Entitled
Honey From A Weed
By Patience Gray

Ye gods what a sight an escaping goat and well he might flee for his life : He sensed the coming feast of sacred cow and unsacred goat (delicious beast!)

To be consumed with ritual, cabbage pots and leek. Wine of course, the whole meal very Greek. The boar, sensing his fate, broke loose of tether and made for hills (hell for leather). But seen by a madcap youth of fine Olympic caste who knew he was equally fast such precision of foot and eye. To noose a goat and cost free training to boot, for games and laurels ahead to prove!
An exhilarating moment, By Jove!

J Fieldhouse 201

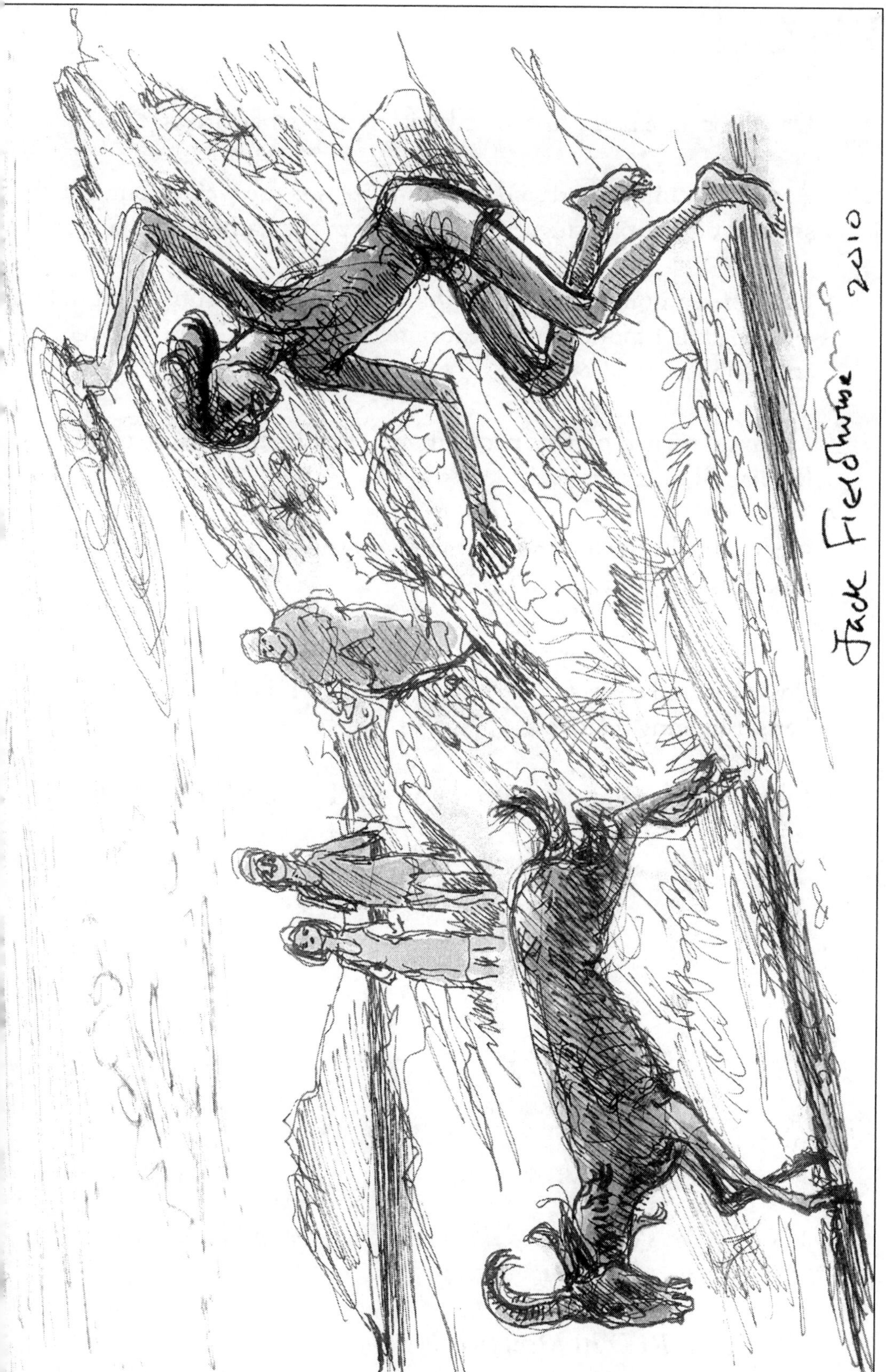
Jack Fieldhouse 2010

Pig Gall

Daughter Julie and I had an exhibition in Nether Stowey.

A woman came in and said how suitable was the wall colour, which showed off the paintings. Could she have the name of the colour?

I searched amongst the tins in the cupboard, paint runs obscured the wording but I found the appropriate tin. It had 'PIG GALL red' still visible.

My mind turned on this peculiar and what seemed to me to be a Mediaeval title.

Hence the illustrator and the man who supplied the disgusting stuff was Edwin Claptrap.

A mind boggling preparation! The flies loved it and M'Lady says, slosh it round them tapestries, the smell and the flies will keep away her scrounging relatives.

Eventually, I found the truth it was PICTURE GALLERY RED!

M'Lady says, slosh it round them tapestries, the smell and the flies will keep away her scrounging relatives.

PIG GALL RED Its Mediaeval beginnings

SWINE WASH
SMEARERS
EDWIN CLAPTRAP

PIG GALL
RED
St. Mediaeval
beginning

M'Lady says, Slosh
round them Tapestries; the smell
and flies keep away 'er scrounging relatives

A Gaggle Of Geese!
No, A Giggle Of Girls

And three cheers for lifting our day
As we entered a coffee house in Brecon, hurray!
Look you now I did not reckon
On seeing these scantiest, flounciest, objects of joy.
Be ribboned, be feathered, a sight to behold.
Unless of course ones cast in a mould or a safe cage from which to
frown.
At the lack of regimental gown to cover all
Some would say eyes down
Avoid such nearly nakedness
To the uncaged lift your eyes
No birds of pray these but Mother Nature's 'critters'

With an unenclosed mien and an unencumbered form to boot fillies
certainlee and just for now a casting off as in snaky slough of excesse
after all it's June and hot and laughter pervades. But mostly giggles a
is their wont and all around are shocked or 'mazed at what they see i
is of course a hen partee

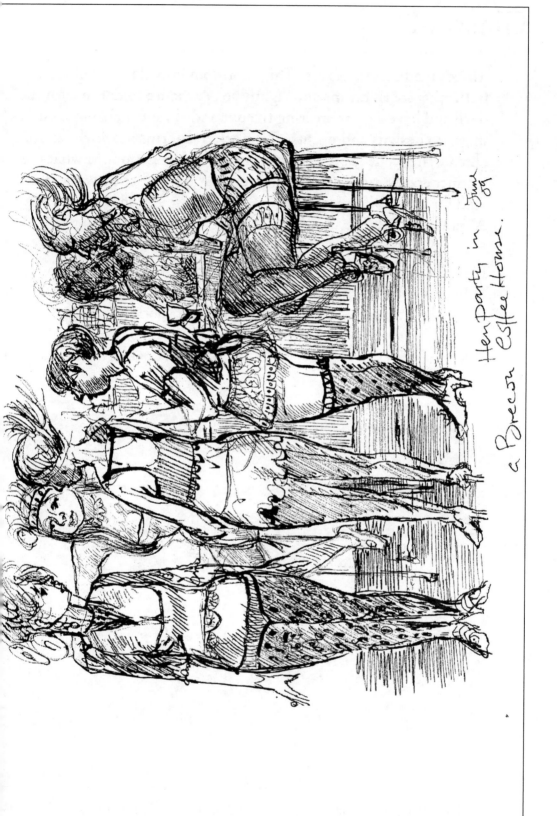

Hen Party in June 09

a Brecon Coffee House.

Glastonbury

There's madness in the air. This enormous crowd has a flair for upturning social norms and rightly so. A chance to dance with the devil and have a go at creating the outlandish, and a chance to shock in garish designs and outfits. To mock the norms in which we are all glued. The scene is pure theatre and a change in mood is what the instinct craves, sobriety no. And just for now its 'let the wine flow.' And for those who recognize a magic star the Pilton buzz word is Abra Cad Ab Ra!

Hail the musack to which a sardine could sympathise. A cramming of bodies with arm exercise to sway with the rhythms at Worthy Farm. So a toast to Mother Nature and her soothing balm.

A triumph for Evis! 201

A 'cider' toast to the health of Mother Nature at Glastonbury
J Fieldhouse June 07

SOMERSET CIDER

'A CIDER' TOAST TO THE HEALTH
OF MOTHER NATURE.
at Glastonbury. J Fieldhouse
June
07

Forever Nagig

How 'portent is the day called Now Twas' ever thus to those who
might endow the day with serious thought: The scene is set for 'UG'
The Mediaeval Chiseller to beget a lasting monument assembled
thus: A sheela; a seat of comfy furs to aid the pose san fuss.

For 'UG' The Chisel (dubbed by locals) whose stone is leaned for ligh
and ease. He the mighty mind who dreamed this day from darkest
winters hours.

Now, UG he can at last apply his arty powers. The muses thus
assembled brings together his naked truthful form 'above and nether'
and now ten thousand 'nows' have passed and some 'disdainful' love
the modern poses with 'a not really very nice' or simply 'Holy Moses'!

But UG has shown the way we come unfiltered with no shade of
gloomy glum Na Gig hides in quiet nook in Musee Llandos –
take a look!

J Fieldhouse April 201

SHEILA NAGIG poses for a sculpture by UG.
J Fieldhouse 2010

SHEILA NAGIG poses for a sculpture by U.G.
Jack Fieldhouse 2010

A Thomas Hood Moment

The illustration was prompted on reading Thomas Hood's Irish
Schoolmaster. It shows the interior of the college of Kilreen.
The pig interrupting the lesson is driven out.

One stanza might help to give the tenor of the place.

For some are meant to right illegal wrongs and some for doctors of
divinity, whom he doth teach to murder the dead tongues and so win
academical degree. But some are bred for service of the sea howbeit,
their store of learning is but small for mickle waste he counteth it
would be to stock a head with bookish wares at all only to be knocked
off by ruthless cannon ball.

M Thomas Hood 'The Irish Schoolmaster' Circa 1850
J Fieldhouse 2009

"THOMAS HOOD CIRCA 1850
"THE IRISH SCHOOLMASTER"
Jack Freshwater 2009

John Betjeman

In a Bath Teashop.

Let us not speak for the love we bear one another – let us hold hands
and look. She such a very ordinary little woman, he, such a thumping
crook. But both, for a moment, little lower than the angels in the
teashops ingle nook.

She such a very ordinary little woman, he, such a thumping crook

(Betjema

She such an ordinary little woman.
He, such a Thumping crook.
(Betjeman)

These quick sketches were

These quick sketches were made during a concert given by the Galliard Players somewhere in Nottinghamshire.

It was Mediaeval music and poetry reading though the reader couldn't resist a poem by Pam Ayres to finish the programme.

I spoke to them afterwards about the quality of sound of their instruments and that the early instruments were varnished with oils containing Propolis.

The violins by Stradivarius were certainly treated thus.

Propolis is a Greek word, the Pro meaning before and Polis in the 'City'.

It was given to the substance, which the bees manufactured from their own bodies it is brown and pliable when warm and is used to fill in cracks where draughts might occur.

In Greece the bees make their homes in a rocky face, it's cool in the summer but come autumn the entrance is reduced to a small hole, the whole of the spacious summer entrance is covered by a thin sheet of this special substance – warm and malleable when produced and quickly hardening after a few hours.

Impossible for predators such as wasps and hornets to enter the nest and so would ensure the survival of the colony into the following year

The early instrument makers warmed this substance Propolis to blend with other oils to apply to the wood and which gave an enhanced 'Timbre' to the sound.

J Fieldhouse C

Jack Fieldhouse 09 The Corboe

Jack Fieldhouse 09

09 Jack Fieldhouse PurAyres' (finale)

Sara

Jack Fieldhouse 09

Jack Fieldhouse 09

Reading Paras Ayres 09 Jack Fieldhons

Sclerda Abbey : The Offertory

Everything light : subtle tones, lovely colours and all a delight to the eye and added to that young curvy forms and hair in a plait or bunched behind, or escaping tresses which sometimes hides from holy guessers or expression : after all it's something new to have a changing face from pew to pew. With a wide smile, a momentary shift of mouth, denoting a juvenile drift of thought and yet again the poise is youth and it certainly annoys some members of the congregation who permanently house an irritation for something duller. (And not the enchanting water colour) for them they prefer an usher in well ironed brown an almost sinister figure and a religious frown, in absolute silence, no squeaky shoes and an air of submit generously to the Godly News. That priestly words proclaim from biblical stews. The old dull picture (ten coats of varnish and designed to re-tarnish) the acceptable and the safe but not up to date. Put your money in the plate! Enjoy the smile, freckles and hair and for gods sake don't be afraid to stare!

Sclerda. Anna+ says they give more to the girls
Anna fists at the Offertory

J Fieldhou

Anna Gets all the attention

"Selecta Jack Fetchouse more to the girls!
Anna says "Then gi'v me more
to the girls!

Anthea Hobnobs With A Goblin

Anthea is a juvenile grandmother and owns a camera, just another excuse for snapping, in a moment of elation and to add to the weight accumulation of moments gone by.

On this occasion there's 'no smile please' just an odd caste of a gnome with inflexible knees and a twisty grimace to boot and impish and ugly, children look aghast at this friend of hobgoblins made out of 'Plast'.

American I suppose and born in a bank non floreata of course.

But has that something of a Yankee swank.

J Fieldhouse Aug 201

Anthea photos the reluctant gnome
J Fieldhouse 2010

Another photo: the reluctant groom

A Side Show At Glastonbury

Into the Marquee de Maeve drifted the Marquis de Sade. Well, someone like him, slightly tipsy, glass in hand, he spies this gorgeous wench just sitting on a bench. The perfect woman quiet and just staring into space. Normally covered in lavender and lace but this one somehow was a bit rummy as he suddenly realised she was a dummy. "Silly" he said, "I must get a grip, well, not on her, but me".

J Fieldhouse Sept 201

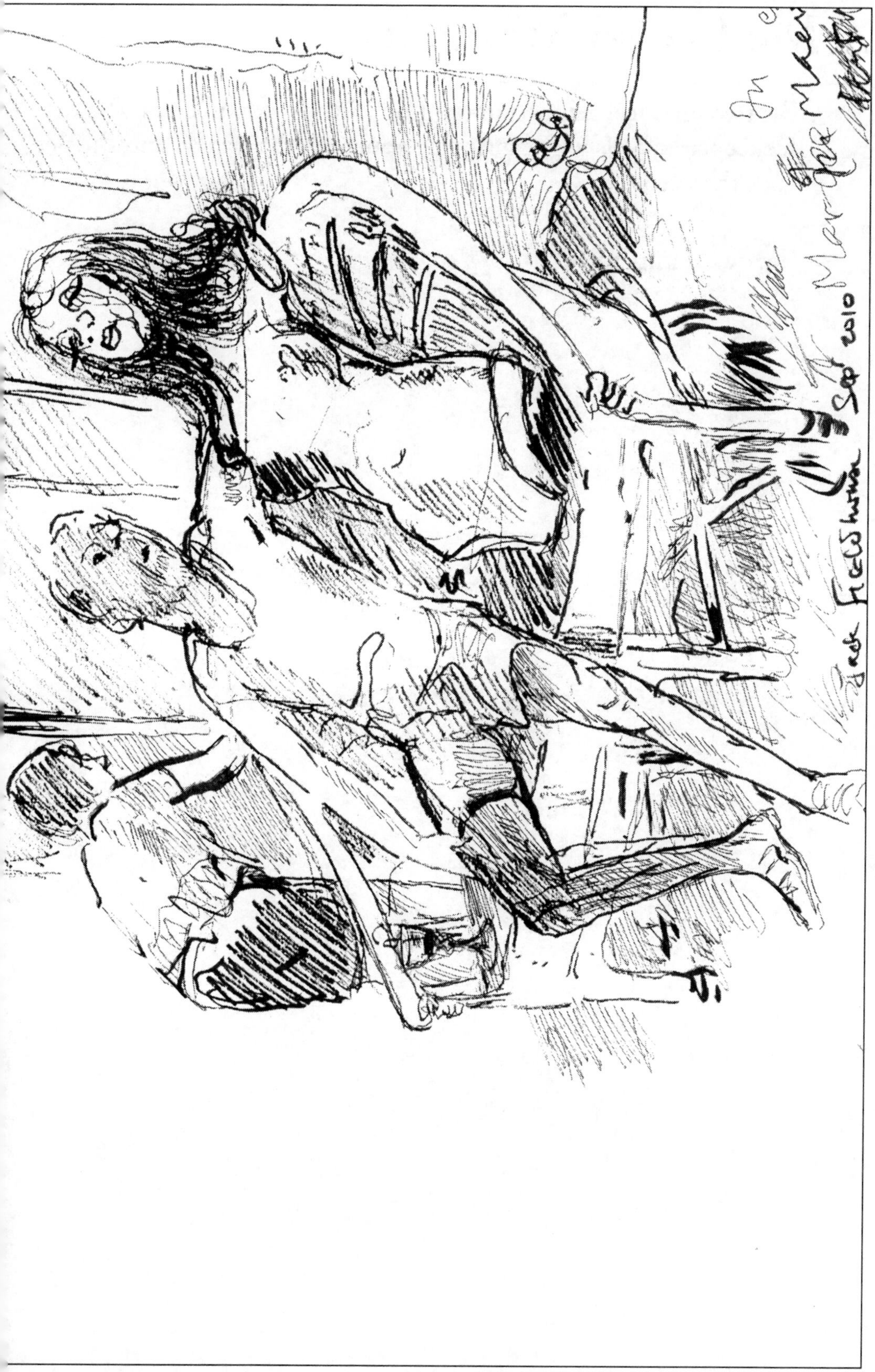

A Mediaeval Moment

The two girls feel trapped in the Loo (inside rotten Tree) The Beekeeper and his wife cautiously approach the swarm and the per stork threatens.

The owl is impeturbed

A Mediaeval Moment!
The Two Suits feel trapped in the Loo.
The Beekeeper and his wife cautiously approach the swarm . .
and the persistic threaten. The Owl is unperturbed
Jack Fieldhouse 64

The Irish Schoolmaster (early 1800's)

The Irish Schoolmaster (early 1800's) is again interrupted by the persistent pig: This 'College of Colleen' was in fact a barn like structure with an earth floor, few windows and in Summer a large open door.

Simple stools, a slate for each with chalk, and the Master with rough coat, a cap and a cane taught children of various ages. The rod was used to keep discipline, the whole scene very primitive.

"THOMAS HOOD CIRCA 1850"
"THE IRISH SCHOOLMASTER"
Jack Fieldhouse 2017

Country Pastimes:
Saying hello to Grannie

The outside Loo was always considered too big, but was built to be a three seater at a time when gossip was at a premium.

Over the years it began to tilt due to the weight of whatever at the rear end.

It was still used by the older generation who had fond memories of extended contemplation and in spite of the toilet paper being cut up bits of The News of the World

Country Pastimes: Saying Hello to Grannie.   Jack Fieldhouse 07

Suddenly an Audience

The scene is set in mucky Rotherham Ely, an allotment keeper and in common with most gardeners always pushed a trolley (home made) and kept an eye out for horse – muck.

In the 30's horses were everywhere. They pulled drays for iron, milk floats, coal delivering and other essentials.

Ely suddenly has an audience!

Men usually had flat caps and for some odd reason wore bicycle clips

A memory moment from 1929

A memory memans from 1929.

Jack Fieldhouse 2012

Exodus from Glastonbury

After 4 days of hilarious romp. The deafening noise of 'musack' mud mud and more mud and the game of sardines while listening to the ear boggling sounds, the moment comes to depart.

Whatever the discomfort – it was worth it.

Exodus from Glastonbury 07

J.Fieldhouse

Josh Swineberry his Goose 'Gess'

Josh Swineberry was an entrepreneur in the early 1800's

His idea of guessing the weight of a goose (and making money) came to him one night when with five other butchers and their wives they hired this great bed of Ware in a Pub called 'The Gander'

It was a celebration of 'Life' it being the week which started with Earth Day (April 22nd)

The locals called it 'Romp' week an old Druid custom of Spring and the reawakening of Nature.

All of course enjoyed by a healthy indifference to the 'officious' Church

The Great Bed of Ware can now be seen in the Victorian and Albert Museum in London and sleeps twelve.

P.S. Joss was thinking of changing his name to Saintsberry because of the numbers of church goers in the village and that it might attract a better class of customer.

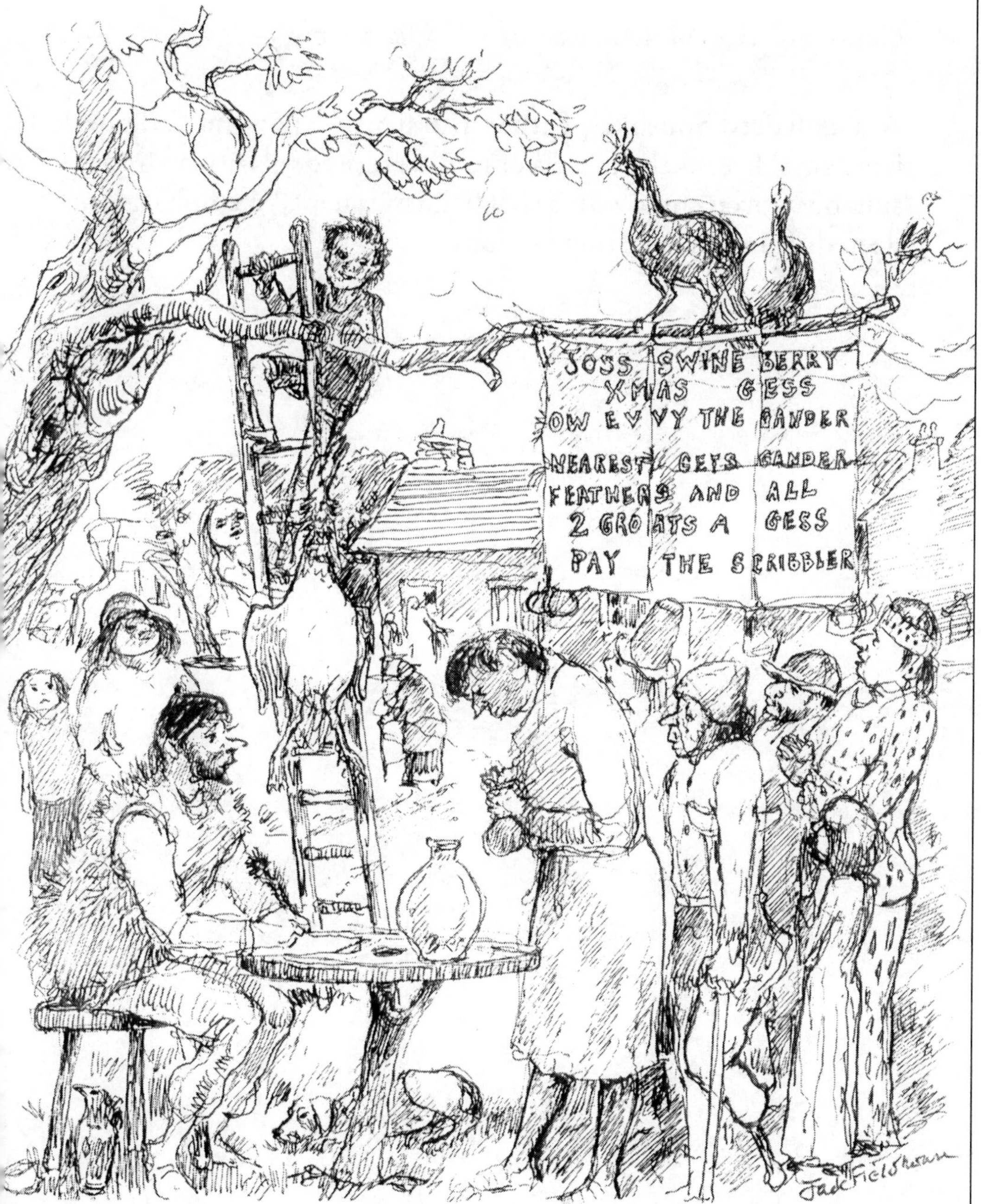
JOSS SWINE BERRY
XMAS GESS
OW EVVY THE GANDER
NEAREST GETS GANDER
FEATHERS AND ALL
2 GROATS A GESS
PAY THE SCRIBBLER

Glastonbury 2011

Contemplative moments are rare at Glastonbury

A dark haired Shushka (Father Russian and how apt and poetical her name) breaks the mould of uninterrupted hilarity by finding this suitably woven chair with a tall haloistic shape, to emphasise her deep drift into meditation and peace.

Pensive Helga
Glastonbury

J Fieldhouse
2011

Sister Clare on her mule

Cleo Lane is a film Artiste and she and a certain Clint Eastwood were in a film entitled "Sister Clare and her two mules"

Their journey through Indian Territory was fraught to say the least, She as a nun was given the respect one would expect for someone who had taken Holy Orders.

An encounter with hostile Indians left poor Clint with an arrow in the shoulder and its subsequent removal was most harrowing

Eventually they this one horse town where our Nun casts odd her habit, joins her friends at the local Brothel and donned the most spectacular outfit – sheer flounce with red dress, a corset of silk. Pearls and tremendous hat of flowers and feathers.

I dived for my sketch book – the spectacle was overwhelming.

And I made a painting of this supreme moment!!

Jack Fieldhouse (from the Telly 2012)

CLEO
LANE
as Sister Clara *[illegible]*
on a Mule *[illegible]*

Jack Fieldsboro 2011

O'Burges boat is not afloat

Lady McFudd never recycled her undies, but gave them willingly to the Kinders of Ork

On this occasion a few years ago they painted a Xmas Pudding on one garment and flagged it for all to see.

O'Burges first boat had too much of a Viking influence and was not manoevrable. It was left on the beach where its stern was flooded.

Little Cloe sails her boat in it

O'Buxyers Boat
is not afloat
We borrit an Undie
from Lady McFudd
And on it we paintit a Krisscimus Pub.
And in weed we wish you cheer
with of course a bottle of beber.

From the
Kinders of ORK

J Fieldhouse 08

On the Isle of Ork 2010

Skipper O'Burge has landed a crate of Saithe and Mackeral.

What to do? Invite the neighbours to a barbeque on the beach

Out comes the half barrel, and with a driftwood fire and 4 lbs of butter the scene was set.

A raging fire is soon acquired and with cardboard plates and bread rolls, the fish are our minutes frying and deft fingers separate bone from flesh and if you want a soft roll dip it in the Barbie – it will probably come up with lots of black carbonny bits attached – but all part of the fun !

Fish cooked 20 at a time and plenty for all.

Finger licking and greasy handkerchiefs complete the meal.

All that is discarded litters the beach and gulls and high tide will do the cleaning.

What a splendid moment

A Burge Barbecue Afternoon of

The Broomfield Buddha

He's a next door Budda

And smaller than a bigga Budda

He occupies a splendid spot

To welcome vizztors, Indian or not

Oh wher's he been this mystic form?

Burn in a clime where warm is norm

In a venerable spot?

Or a bower, not visited a lot?

Where aura pervades of calming chaste

(certainly not to everyones taste)

But now the hard man faces faces

Of Happy children, uncomplicated and ordinary

And refreshingly clean

Mothers too can also be seen

Come to view this copy of a copy

And an original it therefore must be

Though expression 'Tinged with a bit of disnee?'

No matter, he enjoys this childish

Play of head patting, and a turn

Of head to Mummy, What do I say?

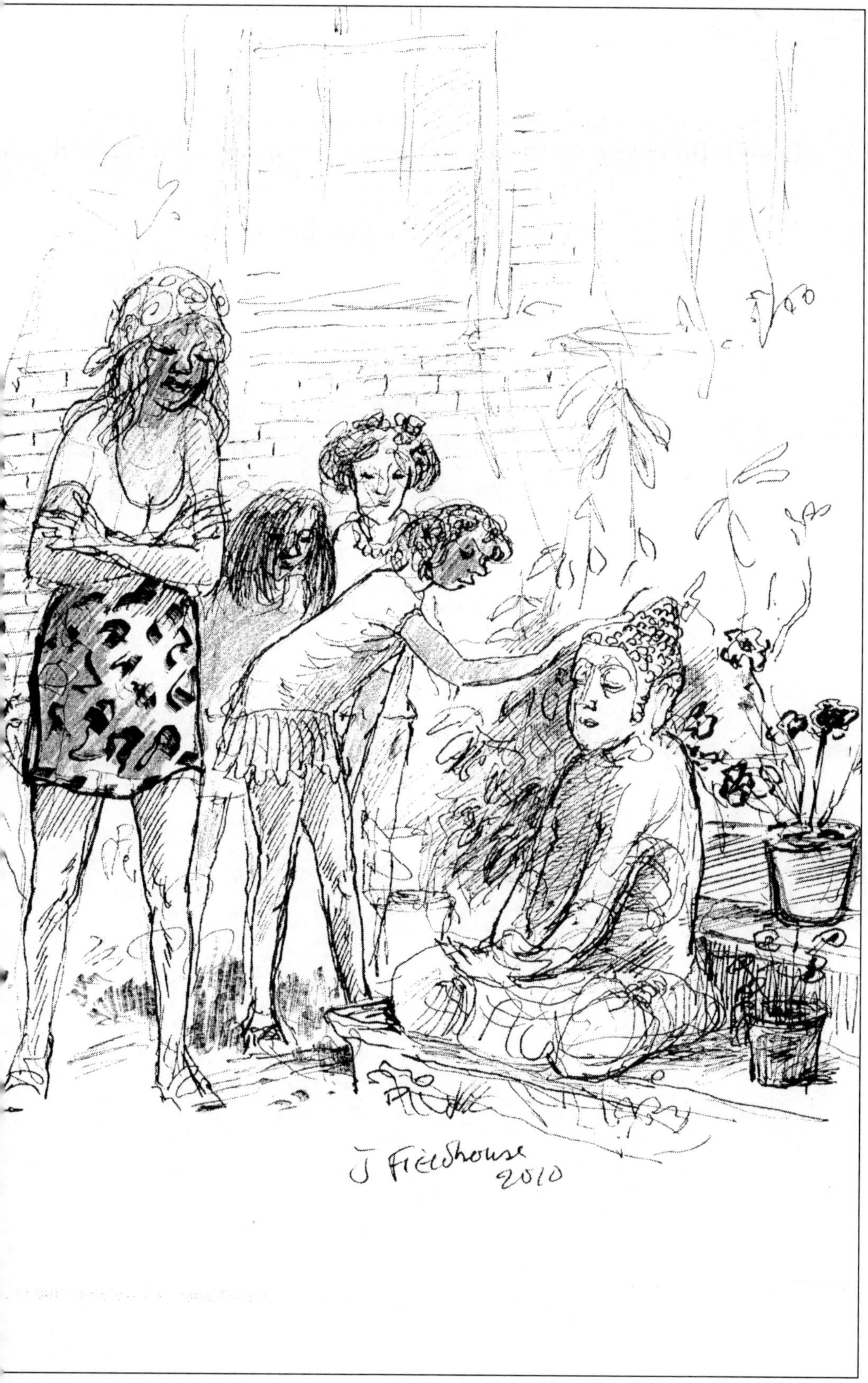
J Fieldhouse
2010

Donkey carrying picture

This is a sketch I made of a Spanish Artist delivering a picture of a famous Bull to an Art Connoisseur & someone who loves a Bull figh

The Donkey is tired and the Driver gives him a rest

Jack FielShorse Sepr 2010

O'Burges early Boat House

This early boat house was erected by O'Burge and his wife Sheelagh, and with the help of friends and neighbours from materials found on the beach.

It wasn't a success as a gale reduced it.

A leaning curve for O'Burge one might say

The Barge Boat House 2011 Jack Fieldhouse

We Waarm T'watter

Five Aunts of mine loved in a large house in St. Annes Road in Rotherham, Yorkshire.

They had a small tarmac'd yard, a wash house detached from the house, and being Yorksgire frugal prepared all the meals on an open gas stove.

Over the wall bordering their house was the graveyard of St. Stephens Church.

Aunt Gertie, the eldest of the family (all Catholics purus) leaned on the wall and asked a woman standing there "why is it that you get more Baptisms in your Church than they do in't Parish Church.

Came the answer "we waarm t'watter!"

Translated in to 'posh' its "we warm the water".

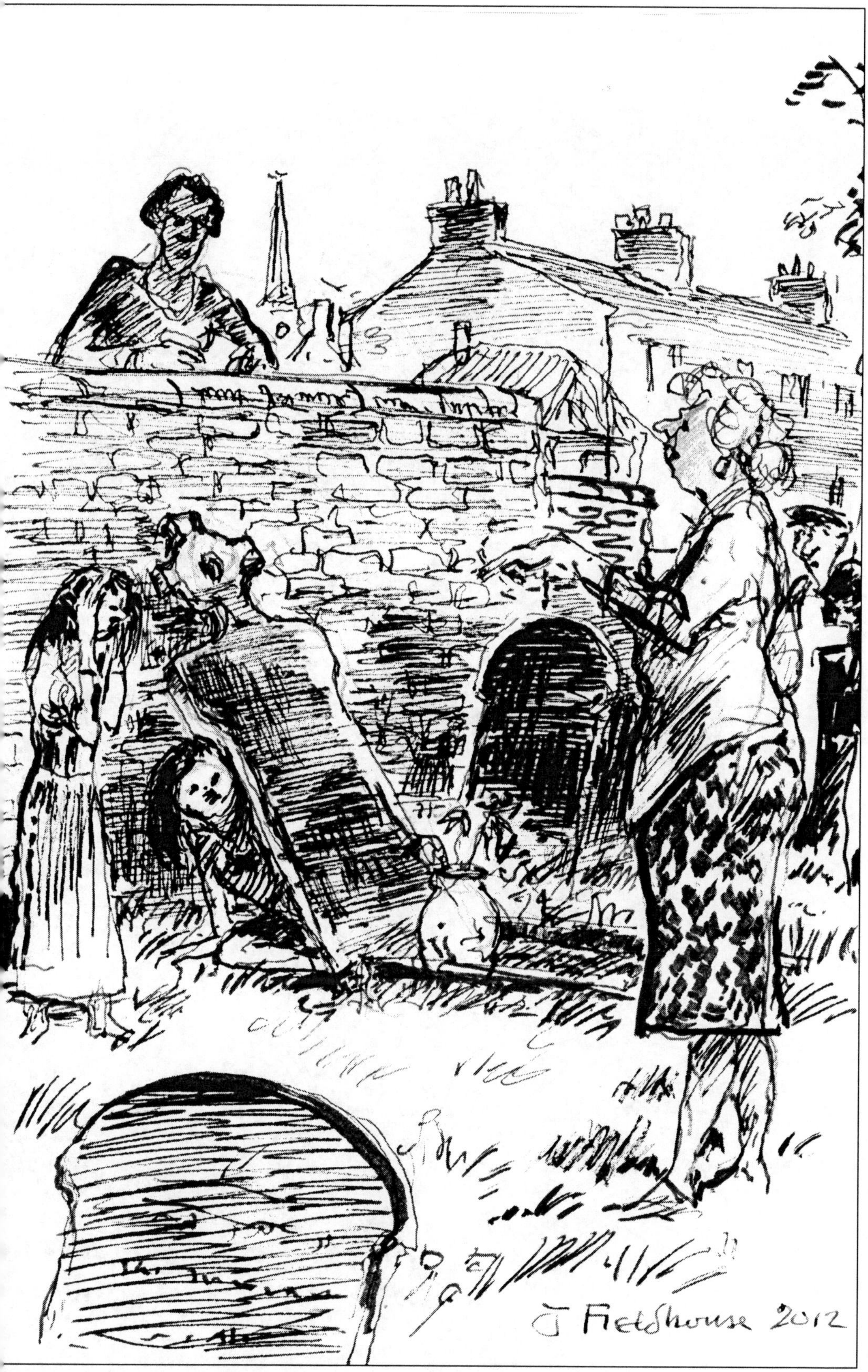
J Fieldhouse 2012

Exodus from Glastonbury

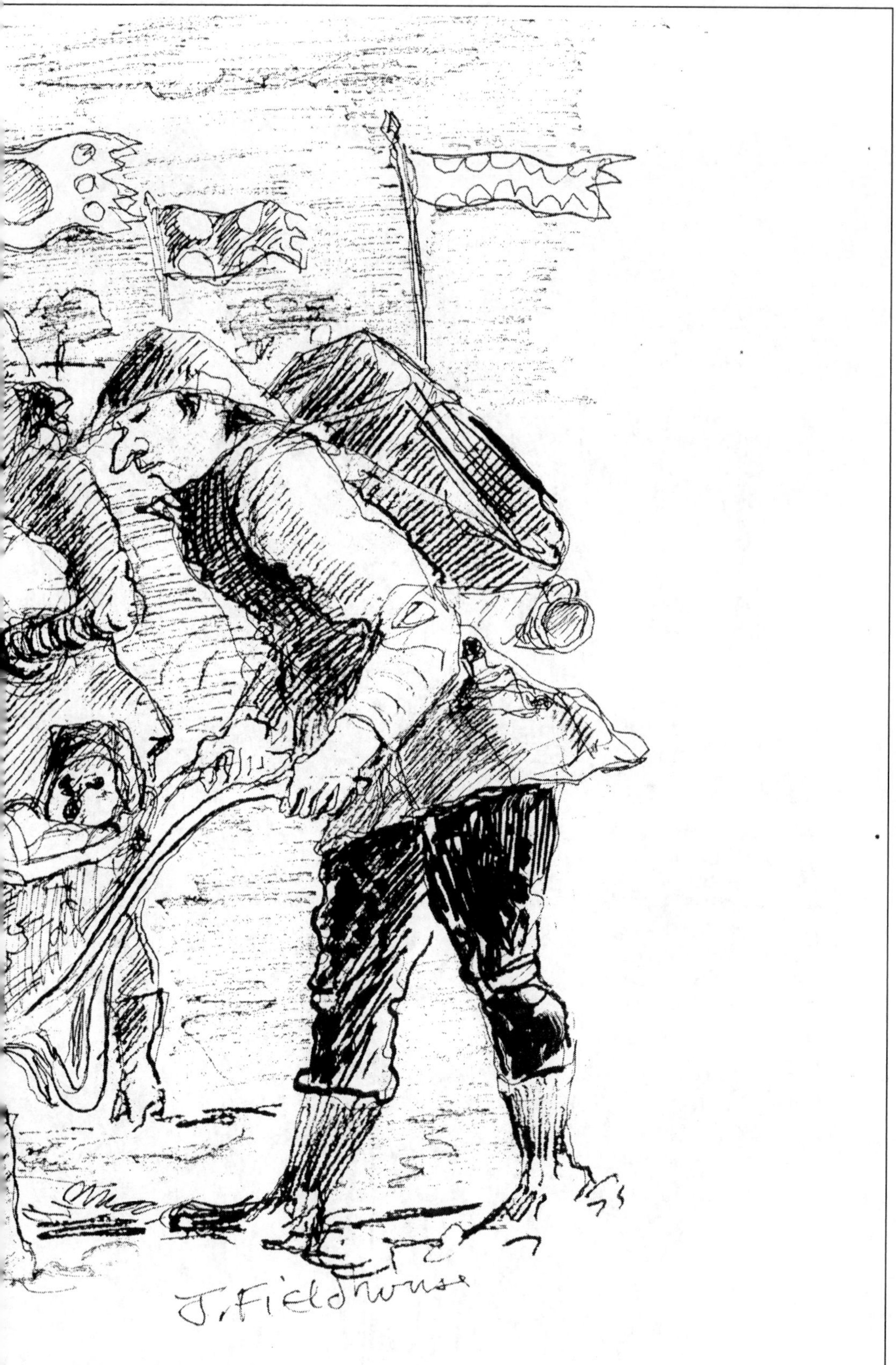
J. Fieldhouse

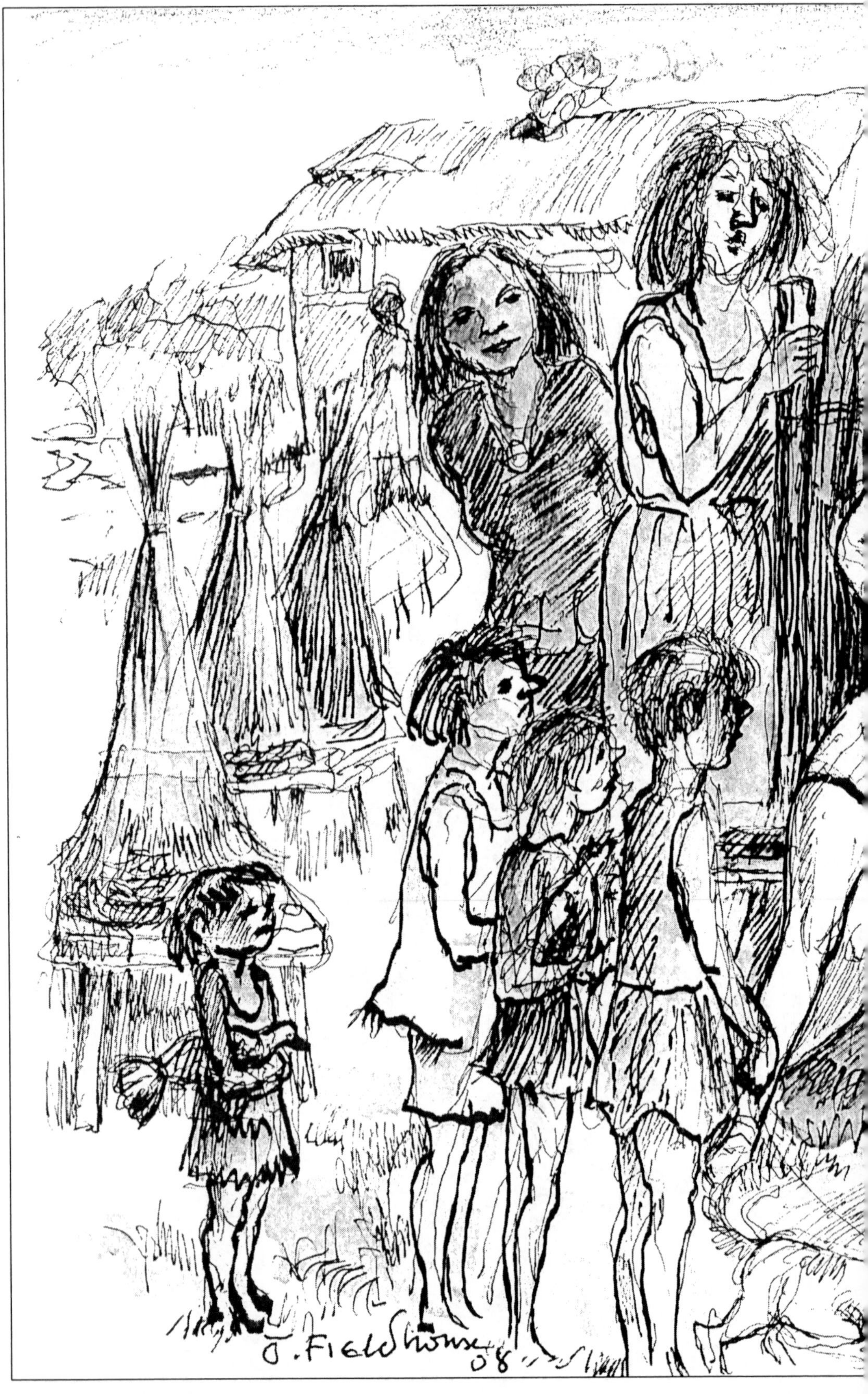
J. Fieldhouse 08

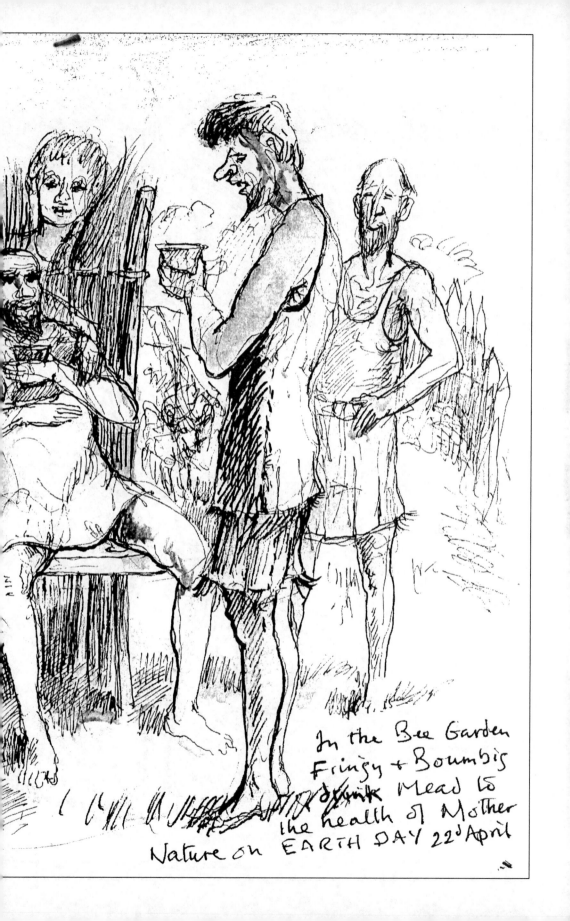

In the Bee Garden
Fringy + Boumbis
drink Mead to
the health of Mother
Nature on EARTH DAY 22d April

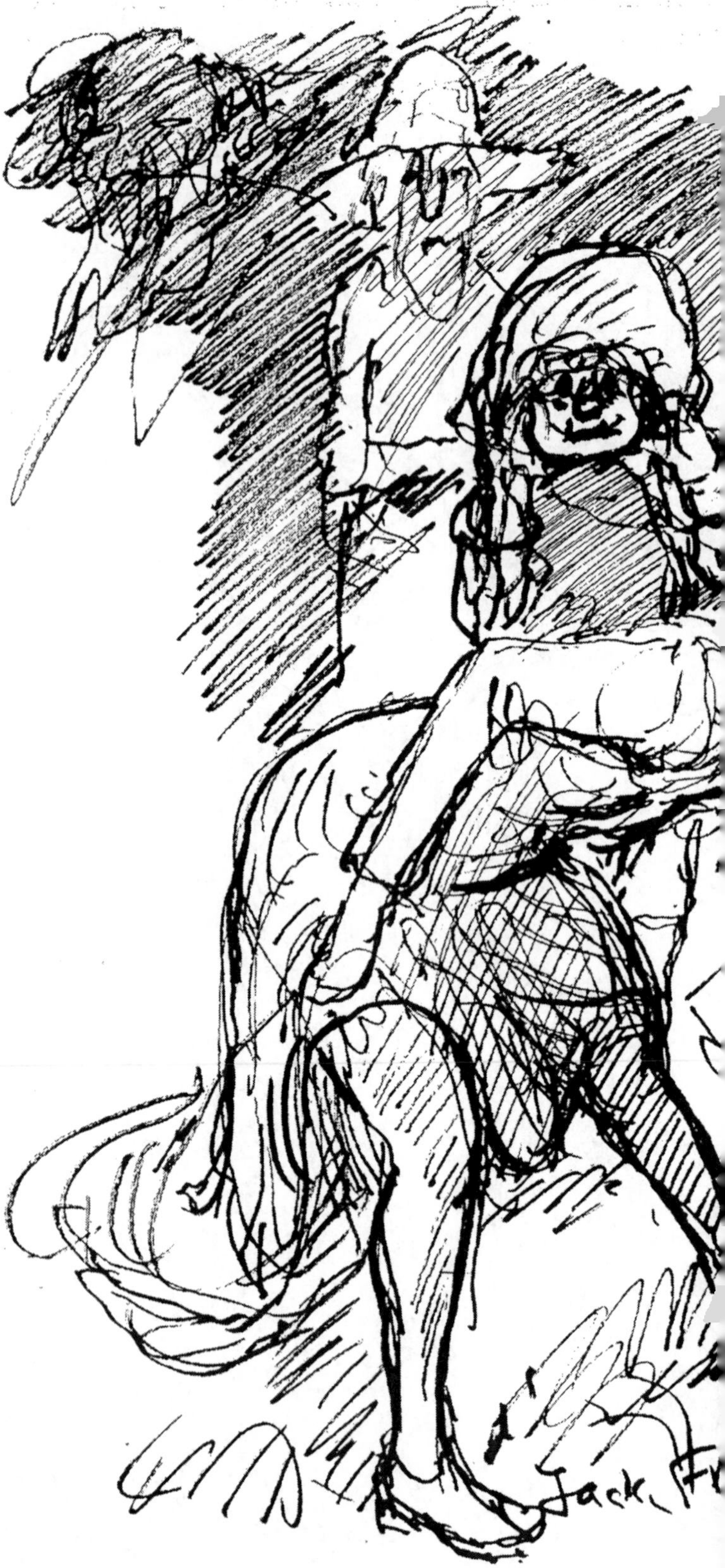

overdressed

The group one with
onlookers only partially
concerned.

Postscript

A bright sunny morning in Italy and the war was over.

As is my want I stray to look around and found five old men playing 'crib' outside a cottage and close to a minefield.

I addressed Mother Nature Thus! You have brought me through unscathed, so more is one test of what you have taught me, and I will walk through that minefield and test my powers of observation.

I conjoured up the thought of Willi Doom Kopt and his mate laying the mines. They would be instructed to make a slight depression, lay the mine and cover with earth carefully.

Carelessness could be fatal and they would be pleased when the job is over.

I trod on undisturbed ground, the disturbed areas I avoided.

In less than ten minutes I was through intact and able to say to my god a thank you, which I know Mother Nature would not hear, but I feel grateful nonetheless.

And I just had two words (I was very good at precis in English Lessons) and it came out as 'Ta Ma' which was enough.

One of my teachers at my Sheffield De La Salle school would no doubt have frowned but it was better and more easily remembered than those long winded prayers inflicted on us by the strap wielding brothers especially the Head Brother Columbus.

Such a good school and splendid discipline!

However the tip toeing iver I looked back at the see the five men resumming their game.

Later quite naturally I pondered on the journey through the maye and think why I had made a decision to risk all.

life I thought was a bit of a minefield and although most of the hazards are glaring obvious there are cunning traps waiting. The ongoing task is to track the mind to recognise such.

Read the wise: Mac Ehrmann in this 'Desiratar' speaks of wholesome discipline and the Great Emperor Hadrian tells us not to be envious of the blind.

And the supreme condensor of thought who poetically arrnages in Quatrains is OMAR KHYAMM:

Brilliantly he writes:

How sweet is mortal sovreighty say some.

Others: How blest the Paradise to come

Ah! Take the cash and wave the rest

How sweet the sound of a distant drum!

Broomfield 2012

Index (Visual Journey)

Lightning Source UK Ltd.
Milton Keynes UK
UKOW02f0715281113

222007UK00001B/4/P

9 781908 90411